W9-BUO-571

The Undoing Project

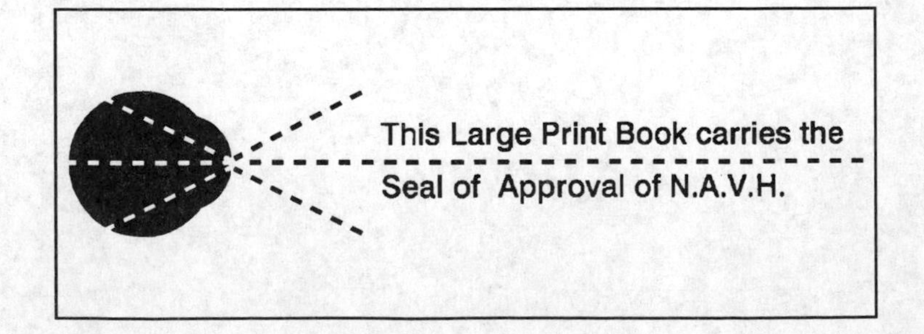
This Large Print Book carries the
Seal of Approval of N.A.V.H.

The Undoing Project

A FRIENDSHIP THAT CHANGED OUR MINDS

Michael Lewis

THORNDIKE PRESS
A part of Gale, Cengage Learning

Farmington Hills, Mich • San Francisco • New York • Waterville, Maine
Meriden, Conn • Mason, Ohio • Chicago

Thorndike Press, a part of Gale, Cengage Learning.

Thorndike Press® Large Print Popular and Narrative Nonfiction.
The text of this Large Print edition is unabridged.
Other aspects of the book may vary from the original edition.
Set in 16 pt. Plantin.

LIBRARY OF CONGRESS CATALOGING-IN-PUBLICATION DATA

Names: Lewis, Michael (Michael M.)
Title: The undoing project : a friendship that changed our minds / by Michael Lewis.
Description: Large print edition. | Waterville, Maine : Thorndike Press, [2017] | Series: Thorndike Press large print popular and narrative nonfiction | Includes bibliographical references.
Identifiers: LCCN 2016047528| ISBN 9781410496454 (hardcover) | ISBN 1410496457 (hardcover)
Subjects: LCSH: Cognitive neuroscience. | Neurosciences. | Decision making. | Statistical decision.
Classification: LCC QP360.5 .L49 2017b | DDC 612.8/233—dc23
LC record available at https://lccn.loc.gov/2016047528

Published in 2017 by arrangement with W. W. Norton & Company, Inc.

Printed in the United States of America
3 4 5 6 7 21 20 19 18 17

For Dacher Keltner
My Chief Jungle Guide

Doubt is not a pleasant condition, but certainty is an absurd one.

— *Voltaire*

CONTENTS

INTRODUCTION: THE PROBLEM THAT NEVER GOES AWAY

Back in 2003 I published a book, called *Moneyball,* about the Oakland Athletics' quest to find new and better ways to value baseball players and evaluate baseball strategies. The team had less money to spend on players than other teams, and so its management, out of necessity, set about rethinking the game. In both new and old baseball data — and the work of people outside the game who had analyzed that data — the Oakland front office discovered what amounted to new baseball knowledge. That knowledge allowed them to run circles around the managements of other baseball teams. They found value in players who had been discarded or overlooked, and folly in much of what passed for baseball wisdom. When the book appeared, some baseball experts — entrenched management, talent scouts, journalists — were upset and dismissive, but a lot of readers found the story as interesting as I had. A lot of people saw in Oakland's approach to building a baseball team a more general lesson: If the highly paid, publicly scrutinized employees of a business that had existed since the 1860s could be misunderstood by their market, who couldn't be? If the market for baseball players was inefficient, what market couldn't be? If a fresh analytical approach had led to the discovery of new knowledge in baseball, was there any sphere of human activity in which it might not do the same?

In the past decade or so, a lot of people have taken the Oakland A's as their role model and set out to use better data, and better analysis of that data, to find market inefficiencies. I've read articles about Moneyball for Education, Moneyball for Movie Studios, Moneyball for Medicare, Moneyball for Golf, Moneyball for Farming, Moneyball for Book Publishing(!), Moneyball for Presidential Campaigns, Moneyball for Government, Moneyball for Bankers, and so on. "All of a sudden we're 'Moneyballing' offensive linemen?" an offensive line coach for the New York Jets complained in 2012. After seeing the diabolically clever data-based approach taken by the North Carolina legislature in writing laws to make it more difficult for African Americans to vote, the comedian John Oliver congratulated the legislators for having "Moneyballed racism."

But the enthusiasm for replacing old-school expertise with new-school data analysis was often shallow. When the data-driven approach to high-stakes decision making did not lead to immediate success — and, occasionally, even when it did — it was open to attack in a way that the old approach to decision making was not. In 2004, after aping Oakland's approach to baseball decision making, the Boston Red Sox won their first World Series in nearly a century. Using the same methods, they won it again in 2007 and 2013. But in 2016, after three disappointing seasons, they announced that they were moving away from the data-based approach and back to one where they relied upon the judgment of baseball experts. ("We have perhaps overly relied on numbers . . . ," said owner John Henry.) The writer Nate Silver for several years enjoyed breathtaking success predicting U.S. presidential election outcomes for the *New York Times,* using an approach to statistics he learned writing about baseball. For the first time in memory, a newspaper seemed to have an edge in calling elections. But then Silver left the *Times,* and failed to predict the rise of

Donald Trump — and his data-driven approach to predicting elections was called into question . . . by the *New York Times*! "Nothing exceeds the value of shoe-leather reporting, given that politics is an essentially human endeavor and therefore can defy prediction and reason," wrote a *Times* columnist, late in the spring of 2016. (Never mind that few shoe-leather reporters saw Trump coming, either, or that Silver later admitted that, because Trump seemed sui generis, he'd allowed an unusual amount of subjectivity to creep into his forecasts.)

I'm sure some of the criticism of people who claim to be using data to find knowledge, and to exploit inefficiencies in their industries, has some truth to it. But whatever it is in the human psyche that the Oakland A's exploited for profit — this hunger for an expert who knows things with certainty, even when certainty is not possible — has a talent for hanging around. It's like a movie monster that's meant to have been killed but is somehow always alive for the final act.

And so, once the dust had settled on the responses to my book, one of them remained more alive and relevant than the others: a review by a pair of academics, then both at the University of Chicago — an economist named Richard Thaler and a law professor named Cass Sunstein. Thaler and Sunstein's piece, which appeared on August 31, 2003, in the *New Republic,* managed to be at once both generous and damning. The reviewers agreed that it was interesting that any market for professional athletes might be so screwed-up that a poor team like the Oakland A's could beat most rich teams simply by exploiting the inefficiencies. But — they went on to say — the author of *Moneyball* did not seem to realize the deeper reason for the inefficiencies in the market for baseball players: They sprang directly from the inner workings of the human mind. The ways in which some baseball expert might misjudge baseball players — the ways in which any expert's judgments might be warped by the expert's own mind — had been described, years ago, by a pair of Israeli psychologists, Daniel Kahneman and Amos Tversky. My book wasn't original. It was simply an illustration of ideas that had been floating around for decades and had yet to be fully appreciated by, among others, me.

That was an understatement. Until that moment I don't believe I'd ever heard of either Kahneman or Tversky, even though one of them had somehow managed to win a Nobel Prize in economics. And I hadn't actually thought much about the psychological aspects of the *Moneyball* story. The market for baseball players was rife with inefficiencies: why? The Oakland front office had talked about "biases" in the marketplace: Foot speed was overrated because it was so easy to see, for instance, and a hitter's ability to draw walks was undervalued in part because walks were so forgettable — they seemed to require the hitter mainly to do nothing at all. Fat or misshapen players were more likely to be undervalued; handsome, fit players were more likely to be overvalued. All of these biases that the Oakland front office talked about I'd found interesting, but I hadn't really pushed further and asked: Where do the biases come from? Why do people have them? I'd set out to tell a story about the way markets worked, or failed to work, especially when they were valuing people. But buried somewhere inside it was another story, one that I'd left unexplored and untold, about the way the human mind worked, or failed to work, when it was forming judgments and making decisions. When faced with uncertainty — about investments or people or anything else — how did it arrive at its conclusions? How did it process evidence — from a baseball game, an earnings report, a trial, a medical examination, or a speed date? What were people's minds doing — even the minds of supposed experts — that led them to the misjudgments that could be exploited for profit by others, who ignored the experts and relied on data?

And how did a pair of Israeli psychologists come to have so much to say about these matters that they more or less anticipated a book about American baseball written decades in the future? What possessed two guys in the Middle East to sit down and figure out what the mind was doing when it tried to judge a baseball player, or an investment, or a presidential candidate? And how on earth does a psychologist win a Nobel Prize in economics? In the answers to those questions, it emerged, there was another story to tell. Here it is.

1
Man Boobs

You never knew what a kid in the interview room might say to jolt you out of your slumber and back to your senses and force you to pay attention. And once you were paying attention, you naturally placed far greater weight on whatever he had just said than you probably should: The most memorable moments in job interviews for the National Basketball Association were hard to consign to some appropriately sized compartment in the brain. In certain cases it was as if the players were trying to screw up your ability to judge them. For instance, when the Houston Rockets interviewer asked one player if he could pass a drug test, the guy had gone wide-eyed and grabbed the table and said, "You mean today!!!???" There was the college player who'd been arrested on charges (subsequently dropped) of domestic violence, and whose agent claimed it had been a simple misunderstanding. When they'd asked the player about it he'd explained, chillingly, that he'd grown weary of his girlfriend's "bitching, so I just put my hands around her neck and I squeezed. 'Cause I needed her to shut up." There was Kenneth Faried, the power forward out of Morehead State. When he showed up for his interview they'd asked him, "Do you prefer to be called Kenneth or Kenny?" "Manimal," Faried said. He wanted to be called Manimal. What did you do with that? Roughly three out of every four of the black American players who came for NBA interviews — or at least came for interviews with the NBA's Houston Rockets — had never really known their father. "It's not uncommon, when you ask these guys who their biggest male influence was, for them to say, 'My mom,' " said the Rockets' director of player personnel, Jimmy Paulis. "One said, 'Obama.' "

Then there was Sean Williams. Back in 2007 Sean Williams, six foot ten, was an off-the-charts player who had been suspended from his Boston College team the first two of his three seasons after being arrested for possession of marijuana (a charge that was later dropped). He'd played only fifteen games his sophomore year and still blocked 75 shots; the fans referred to his college games as The Sean Williams Block Party. Sean Williams looked like a big-time NBA player and was expected to be a first-round pick — in part because everyone assumed that his ability to get through his junior year without being suspended meant that he'd gotten his marijuana use under control. Before the 2007 NBA draft, he'd flown to Houston, at his agent's request, to practice his interviewing skills. The agent cut the Rockets a deal: Williams would talk to the Rockets and the Rockets alone, and the Rockets would offer the agent tips about how to make Sean Williams more persuasive in a job interview. It actually went pretty well, until they got onto the topic of marijuana. "So you got caught smoking weed your freshman and sophomore years," said the Rockets interviewer. "What happened your junior year?" Williams just shook his head and said, "They stopped testing me. And if you're not going to test me, I'm gonna smoke!"

After that, Williams's agent decided it was best for Sean Williams not to grant any more interviews. He still got himself drafted in the first round by the New Jersey Nets, and made brief appearances in 137 NBA games before leaving to play in Turkey.

Millions of dollars were at stake — NBA

players were, on average, by far the highest-paid athletes in all of team sports. The future success of the Houston Rockets was on the line. These young people were hurling information about themselves at you that was meant to help you to make an employment decision. But a lot of times it was hard to know what to do with it.

Rockets interviewer: What do you know about the Houston Rockets?
Player: I know you are in Houston.

Rockets interviewer: Which foot did you hurt?
Player: I have been telling people my right foot.

Player: Coach and I did not see eye to eye.
Rockets interviewer: On what?
Player: Playing time.
Rockets interviewer: What else?
Player: He was shorter.

Ten years of grilling extremely tall people had reinforced in Daryl Morey, the general manager of the Houston Rockets, the sense that he should resist the power of any face-to-face interaction with some other person to influence his judgment. Job interviews were magic shows. He needed to fight whatever he felt during them — especially if he and everyone else in the room felt charmed. Extremely tall people had an unusual capacity to charm. "There's a lot of charming bigs," said Morey. "I don't know if it's like the fat kid on the playground or what." The trouble wasn't the charm but what the charm might mask: addictions, personality disorders, injuries, a deep disinterest in hard work. The bigs could bring you to tears with their story about their love of the game and the hardship they had overcome to play it. "They *all* have a story," said Morey. "I could tell you a story about every guy." And when the story was about perseverance in the face of incredible adversity, as it often was, it was hard not to grow attached to it. It was hard not to use it to create in your mind a clear picture of future NBA success.

But Daryl Morey believed — if he believed in anything — in taking a statistically based approach to decision making. And the most important decision he made was whom to allow onto his basketball team. "Your mind needs to be in a constant state of defense against all this crap that is trying to mislead you," he said. "We're always trying to figure out what's a trick and what's real. Are we seeing a hologram? Is this an illusion?" These interviews belonged on the list of the crap trying to mislead you. "Here's the biggest reason I want to be in every interview," said Morey. "If we pick him, and he has some horrible problem and the owner asks, 'What did he say in the interview when you asked him that question?' and I go, 'I never actually spoke to him before we gave him one point five million dollars,' I get fired."

And so, in the winter of 2015, Morey, along with five members of his staff, sat in a conference room in Houston, Texas, waiting for another giant. The interview room contained nothing worth seeing. A conference table, some chairs, windows obscured by blinds. On the table rested a lone coffee mug, left by mistake, with a logo — National Sarcasm Society: Like We Need Your Support. The giant was . . . well, none of the men knew all that much about him except that he was still only nineteen years old, and that he was huge even by the standards of professional basketball. He'd been discovered five years earlier in a village in Punjab by some agent or talent scout — or so they'd been told. He was then fourteen years old, seven feet tall, and barefoot — or, at any rate, wearing shoes so tattered they revealed his feet.

They'd wondered about that. The kid's family must have been so poor that they couldn't afford to buy him shoes. Or maybe they'd decided it was pointless to buy shoes for feet that grew so rapidly. Or maybe the whole thing was a fiction invented by an agent. Either way, what lingered in the mind was the image: a seven-foot-tall, fourteen-year-old-boy, barefoot in the streets of India. They didn't know how the boy had found his way out of the Indian village. Somebody, probably an agent, had arranged for him to travel to the United States to learn how to speak English and play basketball.

To the NBA he was a complete unknown. There was no video of the guy playing organized basketball. He hadn't played, so far as the Rockets could determine. He hadn't participated in the NBA Draft Combine, the formal audition for amateur players. It was only just that morning that the Rockets had been permitted to take his measurements. His feet were size 22, and his hands, from fingertip to wrist, were eleven and a half inches, the biggest hands the staff had ever measured. Shoeless, he stood seven foot two and weighed three hundred pounds, and his agent claimed he was still growing. He'd spent the past five years in southwest Florida learning basketball — most recently at IMG, a sports academy built to turn amateurs into professionals. Although no one they knew had seen him play, the few people who had laid eyes on him were still talking about it. Robert Upshaw, for instance. Upshaw was a thick seven-foot center who had been dismissed from his team at the University of Washington and was now auditioning for NBA teams. A few days earlier, in the Dallas Mavericks gym, he'd worked out with the Indian giant. Hearing from the Rockets scouts that he might be about to do it again, Upshaw's eyes went wide and his face lit up and he said, "The dude is the biggest human being I've ever seen. And he can shoot the three-ball! It's crazy."

Back in 2006, when he was hired to run the Houston Rockets and figure out who should play pro basketball and who should not, Daryl Morey had been the first of his kind: the basketball nerd king. His job was to replace one form of decision making, which relied upon the intuition of basketball experts, with another, which relied mainly on the analysis of data. He had no serious basketball-playing experience and no interest in passing himself off as a jock or a basketball insider. He'd always been just the way he was, a person who was happier counting than feeling his way through life. As a kid he'd cultivated an interest in using data to make predictions until it became a ruling obsession. "That always seemed the coolest thing to me," he said. "How do you use numbers to predict things? It was like a cool way to use numbers to be better than other people. And I really liked being better than other people." He built forecasting models the way other kids built model airplanes. "It was always sports I was trying to predict. I didn't know what else to apply it to — what, am I going to forecast my grades?"

His interest in sports and statistics had led him, at the age of sixteen, to pick up a book called *The Bill James Historical Baseball Abstract.* Bill James was then busy popularizing an approach, rooted in statistical reasoning, to thinking about baseball. With some help from the Oakland Athletics, that approach would trigger a revolution that ended with nerds running, or helping to run, virtually every team in Major League Baseball. In 1988, when he stumbled upon James's book in a Barnes & Noble, Morey had no way of knowing that people with a gift for using numbers to predict things would overrun professional sports management and everyplace else high-stakes decisions were being made — or that basketball would be, in effect, waiting for him to grow up. He simply suspected that the established experts maybe didn't know as much as everyone thought they did.

That particular suspicion had been born the year before, 1987, after *Sports Illustrated* splashed his favorite baseball team, the Cleveland Indians, on its cover and picked them to win the World Series. "I was like, 'This Is It!!!! The Indians have sucked for years. Now we're going to win the World Series!' " The Indians finished that season with the worst record in the major leagues: How did that happen? "The guys they had said were going to be so good were so bad," recalled Morey. "And *that* was the moment when I thought: Maybe the experts don't know what they're talking about."

Then he discovered Bill James and decided that, like Bill James, he might use numbers to make better predictions than the experts. If he could predict the future performance of professional athletes, he could build winning sports teams, and if he could build winning sports teams . . . well, that's where Daryl Morey's mind came to rest. All he wanted to do in life was to build winning sports teams. The question was: Who'd let him do it? In college he'd sent dozens of letters to professional sports franchises in the hope of being offered some menial job. He received not a single reply. "I didn't have, like, any way to penetrate organized sports," he said. "So I decided at that point that I had to be rich. If I was rich I could just buy a team and run it."

His parents were middle-class midwesterners. He didn't even know any rich people. He was also a distinctly unmotivated student at Northwestern University. He nevertheless set out to make enough money to buy a professional sports team, so that he might make the decisions about who would be on it. "Every week he'd take a sheet of paper and write on top, 'My Goals,' " recalls his then-girlfriend, Ellen, now his wife. "The biggest life goal was, 'I'm going to someday own a professional sports team.' " "I went to business school," said Morey, "because I thought that's where you had to go if you wanted to get rich." Upon leaving business school, in 2000, he interviewed with consulting firms until he found one that got paid in the shares of the companies it advised. The firm was advising Internet companies during the Internet bubble: That sounded, at the time, like a way to get rich quick. Then the bubble burst and all the shares were worthless. "It turns out it was the worst decision ever," said Morey.

From his stint as a consultant he learned something valuable, however. It seemed to him that a big part of a consultant's job was to feign total certainty about uncertain things. In a job interview with McKinsey, they told him that he was not certain enough in his opinions. "And I said it was because I wasn't certain. And they said, 'We're billing clients five hundred grand a year, so you have to be sure of what you are saying.' " The consulting firm that eventually hired him was forever asking him to exhibit confidence when, in his view, confidence was a sign of fraudulence. They'd asked him to forecast the price of oil for clients, for instance. "And then we would go to our clients and tell them we could predict the price of oil. No one can predict the price of oil. It was basically nonsense."

A lot of what people did and said when they "predicted" things, Morey now realized, was phony: pretending to know things rather than actually knowing things. There were a great many interesting questions in the world to which the only honest answer was, "It's impossible to know for sure." "What will the price of oil be in ten years?" was such a question. That didn't mean you gave up trying to find an answer; you just couched that answer in probabilistic terms.

Later, when basketball scouts came to him looking for jobs, the trait he looked for was some awareness that they were seeking answers to questions with no certain answers — that they were inherently fallible. "I always ask them, 'Who did you miss?' " he said. Which future superstar had they written off, or which future bust had they fallen in love with? "If they don't give me a good one, I'm like, 'Fuck 'em.' "

By a stroke of luck, the consulting firm Morey worked for was asked to perform some analysis for a group trying to buy the Boston Red Sox. When that group failed in its bid to buy a professional baseball team, it went out and bought a professional basketball team, the Boston Celtics. In 2001 they asked Morey to quit his job consulting and come to work for the Celtics, where "they gave me the most difficult problems to figure out." He helped hire new management, then helped to figure out how to price tickets, and, finally, inevitably, was asked to work on the problem of whom to select in the NBA draft. "How will that nineteen-year-old perform in the NBA?" was like "Where will the price of oil be in ten years?" A perfect answer didn't exist, but statistics could get you to some answer that was at least a bit better than simply guessing.

Morey already had a crude statistical model to evaluate amateur players. He'd built it on his own, just for fun. In 2003 the Celtics had encouraged him to use it to pick a player at the tail end of the draft — the 56th pick, when the players seldom amount to anything. And thus Brandon Hunter, an obscure power forward out of Ohio University, became the first player picked by an equation.* Two years later Morey got a call from a headhunter who said that the Houston Rockets were looking for a new general manager. "She said they were looking for a Moneyball type," recalled Morey.

The Rockets' owner, Leslie Alexander, had grown frustrated with the gut instincts of his basketball experts. "The decision making wasn't that good," Alexander said. "It wasn't precise. We now have all this data. And we have computers that can analyze that data. And I wanted to use that data in

* Hunter actually started for the Celtics for a season and went on to a successful career in Europe.

a progressive way. When I hired Daryl, it was because I wanted somebody that was doing more than just looking at players in the normal way. I mean, I'm not even sure we're playing the game the right way." The more the players got paid, the more costly to him the sloppy decisions. He thought that Morey's analytical approach might give him an edge in the market for high-priced talent, and he was sufficiently indifferent to public opinion to give it a whirl. ("Who cares what other people think?" says Alexander. "It's not their team.") In his own job interview, Morey was reassured by Alexander's social fearlessness, and the spirit in which he operated. "He asked me, 'What religion are you?' I remember thinking, *I don't think you're supposed to ask me that.* I answered it vaguely, and I think I was saying my family were Episcopalians and Lutherans when he stops me and says, 'Just tell me you don't believe any of that shit.' "

Alexander's indifference to public opinion turned out to come in handy. Learning that a thirty-three-year-old geek had been hired to run the Houston Rockets, fans and basketball insiders were at best bemused and at worst hostile. The local Houston radio guys instantly gave him a nickname: Deep Blue. "There's an intense feeling among basketball people that I don't belong," said Morey. "They remain silent during periods of success and pop up when they sense weakness." In his decade in charge, the Rockets have had the third-best record of the thirty teams in the NBA, behind the San Antonio Spurs and the Dallas Mavericks, and have appeared in the playoffs more than all but four teams. They've never had a losing season. The people most upset by Morey's presence had no choice at times but to go after him in moments of strength. In the spring of 2015, as the Rockets, with the second-best record in the NBA, headed into the Western Conference Finals against the Golden State Warriors, the former NBA All-Star and current TV analyst Charles Barkley went off on a four-minute tirade about Morey during what was meant to be a halftime analysis of a game. ". . . I'm not worried about Daryl Morey. He's one of those idiots who believe in analytics. . . . I've always believed analytics was crap. . . . Listen, I wouldn't know Daryl Morey if he walked in this room right now. . . . The NBA is about talent. All these guys who run these organizations who talk about analytics, they have one thing in common: They're a bunch of guys who ain't never played the game, and they never got the girls in high school and they just want to get in the game."

There'd been a lot more stuff just like that. People who didn't know Daryl Morey assumed that because he had set out to intellectualize basketball he must also be a know-it-all. In his approach to the world he was exactly the opposite. He had a diffidence about him — an understanding of how hard it is to know anything for sure. The closest he came to certainty was in his approach to making decisions. He never simply went with his first thought. He suggested a new definition of the nerd: a person who knows his own mind well enough to mistrust it.

One of the first things Morey did after he arrived in Houston — and, to him, the most important — was to install his statistical model for predicting the future performance of basketball players. The model was also a tool for the acquisition of basketball knowledge. "Knowledge is literally prediction," said Morey. "Knowledge is anything that increases your ability to predict the outcome. Literally everything you do you're trying to predict the right thing. Most people just do it subconsciously." A model allowed you to explore the attributes in an amateur basketball player that led to professional success, and determine how much weight should be given to each. Once you had a database of thousands of former players, you could search for more general correlations between their performance in college and their professional careers. Obviously their performance statistics told you something about them. But which ones? You might believe — many then did — that the most important thing a basketball player did was to score points. That opinion could now be tested: Did an ability to score points in college predict NBA success? No, was the short answer. From early versions of his model, Morey knew that the traditional counting statistics — points, rebounds, and assists per game — could be wildly misleading. It was possible for a player to score a lot of points and hurt his team, just as it was possible for a player to score very little and be a huge asset. "Just having the model, without any human opinion at all, forces you to ask the right questions," said Morey. "Why is someone ranked so high by scouts when the model has him ranked low? Why is someone ranked so low by scouts when the model has him ranked high?"

He didn't think of his model as "the right answer" so much as "a better answer." Nor was he so naive as to think that the model would pick players all by itself. The model obviously needed to be checked and watched — mainly because there was information that the model wouldn't be privy to. If the player had broken his neck the night before the NBA draft, for instance, it would be nice to know. But if you had asked Daryl Morey in 2006 to choose between his model and a roomful of basketball scouts, he'd have taken his model.

That counted as original, in 2006. Morey could see that no one else was using a model to judge basketball players — no one had bothered to acquire the information needed by any model. To get any stats at all, he'd had to send people to the offices of the National Collegiate Athletic Association (NCAA), in Indianapolis, to photocopy box scores of every college game over the past twenty years, then enter all that data by hand into his system. Any theory about basketball players had to be tested on a database of players. They now had a twenty-year history of college players. The new database allowed you to compare players to similar players from the past, and see if there were any general lessons to be learned.

A lot of what the Houston Rockets did sounds simple and obvious now: In spirit, it is the same approach taken by algorithmic

Wall Street traders, U.S. presidential campaign managers, and every company trying to use what you do on the Internet to predict what you might buy or look at. There was nothing simple or obvious about it in 2006. There was much information Morey's model needed that simply was not available. The Rockets began to gather their own original data by measuring things on a basketball court that had previously gone unmeasured. Instead of knowing the number of rebounds a player had, for instance, they began to count the number of genuine opportunities for rebounds he'd had and, of those, how many he had snagged. They tracked the scoring in the game when a given player was on the court, compared to when he was on the bench. Points and rebounds and steals *per game* were not very useful; but points and rebounds and steals *per minute* had value. Scoring 15 points a game obviously meant less if you had played the entire game than if you had played half of it. It was also possible to back out from the box scores the pace at which various college teams played — how often they went up and down the court. Adjusting a college player's stats for his team's pace of play was telling. Points and rebounds meant one thing when the team took 150 shots a game and something different when it took just 75. Just adjusting for pace gave you a clearer picture of what any given player had accomplished than the conventional view did.

The Rockets collected data on basketball players that hadn't ever been collected before, and not just basketball data. They gathered information on the players' lives and looked for patterns in it. Did it help a player to have two parents in his life? Was it an advantage to be left-handed? Did players with strong college coaches tend to do better in the NBA? Did it help if a player had a former NBA player in his lineage? Did it matter if he had transferred from junior college? If his college coach played zone defense? If he had played multiple positions in college? Did it matter how much weight a player could bench-press? "Almost everything we looked at was nonpredictive," says Morey. But not everything. Rebounds per minute were useful in predicting the future success of big guys. Steals per minute told you something about the small ones. It didn't matter so much how tall a player was as how high he could reach with his hands — his length rather than his height.

The model's first road test came in 2007. (The Rockets had traded their picks in 2006.) Here was the chance to test a dispassionate, unsentimental, evidence-based approach against the felt experience of an entire industry. That year, the Rockets held the 26th and the 31st picks in the NBA draft. According to Morey's model, the odds of getting a good NBA player with those picks were, respectively, 8 percent and 5 percent. The chance of getting a starter was roughly one in a hundred. They selected Aaron Brooks and Carl Landry, both of whom became NBA starters. It was an incredibly rich haul.* "That lulled us to sleep," said Morey. He knew that his model was, at best, only slightly less flawed than the human beings who had rendered the judgments about job applicants since time began. He knew that he suffered from a serious dearth of good data. "You have some information — but often from a single year in college. And even that has problems with

* There's no perfect way to measure the quality of a draft choice, but there's a sensible one: comparing the player's output in his first four years, the years the NBA team that drafts him also controls him, to the average output of players drafted in that slot. By that measure, Carl Landry and Aaron Brooks were the 35th and 55th best picks of the six hundred or so picks made by NBA teams in the last decade.

it. Apart from it's a different game, with different coaches, different levels of competition — the players are twenty years old. *They* don't know who they are. So how are we supposed to?" He knew all this and yet he thought maybe they had figured something out. Then came 2008.

That year the Rockets had the 25th pick in the draft and used it to pick a big guy from the University of Memphis named Joey Dorsey. In his job interview, Dorsey had been funny and likable and charming — he'd said when he was done playing basketball he intended to explore a second career as a porn star. After he was drafted, Dorsey was sent to Santa Cruz to play in an exhibition game against other newly drafted players. Morey went to go see him. "The first game I watch he looks terrible," said Morey. "And I'm like, 'Fuck!!!!' " Joey Dorsey was so bad that Daryl Morey could not believe he was watching the guy he'd drafted. Perhaps, Morey thought, he wasn't taking the exhibition seriously. "I meet with him. We have a two-hour lunch." Morey gave Dorsey a long talk about the importance of playing with intensity, and making a good impression, and so on. "I think he's going to come out the next game with his hair on fire. And he comes out and sucks the next game, too." Fairly quickly, Morey saw he had a bigger problem than Joey Dorsey. The problem was his model. "Joey Dorsey was a model superstar. The model said that he was like a can't-miss. His signal was super, super high."

That same year, the model had dismissed as unworthy of serious consideration a freshman center at Texas A&M named DeAndre Jordan. Never mind that every other team in the NBA, using more conventional scouting tools, passed him over at least once, or that Jordan wasn't taken until 35th pick of the draft, by the Los Angeles Clippers. As quickly as Joey Dorsey established himself as a bust, DeAndre Jordan established himself as a dominant NBA center and the second-best player in the entire draft class after Russell Westbrook.*

This sort of thing happened every year to some NBA team, and usually to all of them. Every year there were great players the scouts missed, and every year highly re-

* Before the 2015 season, DeAndre Jordan signed a four-year contract with the Clippers that guaranteed him $87,616,050, then the NBA's maximum salary. Joey Dorsey signed a one-year deal for $650,000 with Galatasaray Liv Hospital of the Turkish Basketball League.

garded players went bust. Morey didn't think his model was perfect, but he also couldn't believe that it could be so drastically wrong. Knowledge was prediction: If you couldn't predict such a glaringly obvious thing as the failure of Joey Dorsey or the success of DeAndre Jordan, how much did you know? His entire life had been shaped by this single, tantalizing idea: He could use numbers to make better predictions. The plausibility of that idea was now in question. "I'd missed something," said Morey. "What I missed were the limitations of the model."

His first mistake, he decided, was to have paid insufficient attention to Joey Dorsey's age. "He was insanely old," says Morey. "He was twenty-four years old when we drafted him." Dorsey's college career was impressive because he was so much older than the people he played against. He'd been, in effect, beating up on little kids. Raising the weight the model placed on a player's age flagged Dorsey as a weak NBA prospect; more tellingly, it improved the model's judgments about nearly all of the players in the database. For that matter, Morey realized, there existed an entire class of college basketball player who played far better against weak opponents than against strong ones. Basketball bullies. The model could account for that, too, by assigning greater weight to games played against strong opponents than against weak ones. That also improved the model.

Morey could see — or thought he could see — how the model had been fooled by Joey Dorsey. Its blindness to the value of DeAndre Jordan was far more troubling. The kid had played a single year of college basketball, not very effectively. It turned out that he had been a sensational high school player, had hated his college coach, and didn't even want to be in school. How could any model predict the future of a player who had intentionally failed? It was impossible to see Jordan's future in his college stats, and, at the time, there were no useful high school basketball statistics. So long as it relied almost exclusively on performance statistics, the model would *always* miss DeAndre Jordan. The only way to see him, it seemed, was with the eyes of an old-fashioned basketball expert. As it happens, Jordan had grown up in Houston under the eyes of Rockets scouts, and one of those scouts had wanted to draft him on the strength of what appeared to him undeniable physical talent. One of his scouts had seen what his model had missed!

Morey — being Morey — had actually tested whether there were any patterns in the predictions made by his staff. He'd hired most of them and thought they were great, and yet there was no evidence any one of them was any better than the other, or the market, at predicting who would make it in the NBA and who would not. If there was any such thing as a basketball expert who could identify future NBA talent, he hadn't found him. He certainly didn't think that he was one. "Weighting my personal intuition more heavily did not cross my mind," he said. "I trust my gut very low. I just think there's a lot of evidence that gut instincts aren't very good."

In the end, he decided that the Rockets needed to reduce to data, and subject to analysis, a lot of stuff that had never before been seriously analyzed: physical traits. They needed to know not just how high a player jumped but how quickly he left the earth — how fast his muscles took him into the air. They needed to measure not just the speed of the player but the quickness of his first two steps. That is, they needed to be even more geeky than they already were. "When things go wrong, that's what people do," said Morey. "They go back to the habits that succeeded in the past. My thing was:

Let's go back to first principles. If these physical tools are going to matter, let's test them more rigorously than they've ever been tested before. The weights we placed on production in college had to go down, and the weights we placed on raw physical abilities had to go up."

But once you started to talk about a guy's body and what it might or might not be able to do on an NBA court, there was a limit to the usefulness of even the objective, measurable information. You needed, or seemed to need, experts to look at the tools in action and judge how well they would function playing a different game, against better competition. You needed scouts to rate a player's ability to do the various things they knew were most important to do on a basketball court: shooting, finishing, getting to the rim, offensive rebounding, and so on. You needed *experts*. The limits of any model invited human judgment back into the decision-making process — whether it helped or not.

And thus began a process of Morey trying as hard as he'd ever tried at anything in his life to blend subjective human judgment with his model. The trick wasn't just to build a better model. It was to listen both to it and to the scouts at the same time.

"You have to figure out what the model is good and bad at, and what humans are good and bad at," said Morey. Humans sometimes had access to information that the model did not, for instance. Models were bad at knowing that DeAndre Jordan sucked his freshman year in college because he wasn't trying. Humans were bad at . . . well, that was the subject Daryl Morey now needed to study more directly.

Freshly exposed to the human mind, Morey couldn't help but notice how strangely it operated. When it opened itself to information that might be useful in evaluating an amateur basketball player, it also opened itself to being fooled by the very illusions that had made the model such a valuable tool in the first place. For instance, in the 2007 draft there had been a player his model really liked: Marc Gasol. Gasol was twenty-two years old, a seven-foot-one center playing in Europe. The scouts had found a photograph of him shirtless. He was pudgy and baby-faced and had these jiggly pecs. The Rockets staff had given Marc Gasol a nickname: Man Boobs. Man Boobs this and Man Boobs that. "That was my first draft in charge and I wasn't so brave," said Morey. He allowed the general ridicule of Marc Gasol's body to drown out his model's optimism about Gasol's basketball future, and so instead of arguing with his staff, he watched the Memphis Grizzlies take Gasol with the 48th pick of the draft. The odds of getting an All-Star with the 48th pick in the draft were well below one in a hundred. The 48th pick of the draft basically never even yielded a useful NBA bench player, but already Marc Gasol was proving to be a giant exception.* The label they'd stuck on him clearly had affected how they valued him: names mattered. "I made a new rule right then," said Morey. "I banned nicknames."

All of a sudden he was right back in the mess he and his model had been hired to eliminate. If he could never completely remove the human mind from his decision-making process, Daryl Morey had at least to be alive to its vulnerabilities. He now saw these everywhere he turned. One example: Before the draft, the Rockets would bring a player in with other players and put him through his paces on the court. How could

* Gasol became a two-time All-Star (2012, 2015) and, by Houston's reckoning, the third-best pick made by the entire NBA over the past decade, after Kevin Durant and Blake Griffin.

you deny yourself the chance to watch him play? But while it was interesting for his talent evaluators to see a player in action, it was also, Morey began to realize, risky. A great shooter might have an off day; a great rebounder might get pushed around. If you were going to let everyone watch and judge, you also had to teach them not to place too much weight on what they were seeing. (Then why were they watching in the first place?) If a guy was a 90 percent free-throw shooter in college, for instance, it really didn't matter if he missed six free throws in a row during the private workout.

Morey leaned on his staff to pay attention to the workouts but not allow whatever they saw to replace what they knew to be true. Still, a lot of people found it very hard to ignore the evidence of their own eyes. A few found the effort almost painful, as if they were being strapped to the mast to listen to the Sirens' song. One day a scout came to Morey and said, "Daryl, I've done this long enough. I think we should stop having these workouts. Please, just stop doing them." Morey said, Just try to keep what you are seeing in perspective. Just weight it really low. "And he says, 'Daryl, I just can't do it.' It's like a guy addicted to crack," Morey said. "He can't even get near it without it hurting him."

Soon Morey noticed something else: A scout watching a player tended to form a near-instant impression, around which all other data tended to organize itself. "Confirmation bias," he'd heard this called. The human mind was just bad at seeing things it did not expect to see, and a bit too eager to see what it expected to see. "Confirmation bias is the most insidious because you don't even realize it is happening," he said. A scout would settle on an opinion about a player and then arrange the evidence to support that opinion. "The classic thing," said Morey, "and this happens *all* the time with guys: If you don't like a prospect, you say he has no position. If you like him, you say he's multipositional. If you like a player, you compare his body to someone good. If you don't like him, you compare him to someone who sucks." Whatever prejudice a person brought to the business of selecting amateur players he tended to preserve, even when it served him badly, because he was always looking to have that prejudice confirmed. The problem was magnified by the tendency of talent evaluators — Morey included — to favor players who reminded them of their younger selves. "My playing career is *so* irrelevant to my career," he said.

"And still I like guys who beat the shit out of people and cheat the rules and are nasty. Bill Laimbeer types. Because that's how I played." You saw someone who reminded you of you, and then you looked for the reasons why you liked him.

The mere fact that a player physically resembled some currently successful player could be misleading. A decade ago a six-foot-two-inch, light-skinned, mixed-race guy who had gone unnoticed by major colleges in high school and so played for some obscure tiny college, and whose main talent was long-range shooting, would have had no obvious appeal. The type didn't exist in the NBA — at least not as a raging success. Then Stephen Curry came along and set the NBA on fire, led the Golden State Warriors to an NBA championship, and was everyone's most valuable player. Suddenly — just like that — all these sharp-shooting mixed-race guards were turning up for NBA job interviews and claiming that their game was a lot like Stephen Curry's; and they were more likely to get drafted because of the resemblance.* "For five years after we

* In 2015 Tyler Harvey, a shooting guard out of Eastern Washington, made the rounds. When asked whose game his most resembled, he said,

drafted Aaron Brooks, we saw so many kids who compared themselves to Aaron. Because there are so many little guards." Morey's solution was to forbid all intraracial comparison. "We've said, 'If you want to compare this player to another player, you can only do it if they are a different race.' " If the player in question was African American, for instance, the talent evaluator was only allowed to argue that "he is like so-and-so" if so-and-so was white or Asian or Hispanic or Inuit or anything other than black. A funny thing happened when you forced people to cross racial lines in their minds: They ceased to see analogies. Their minds resisted the leap. "You just don't see it," said Morey.

Maybe the mind's best trick of all was to lead its owner to a feeling of certainty about inherently uncertain things. Over and again in the draft you saw these crystal-clear

---

"To be honest with you, I'm most like Steph Curry," and he would go on to say that, as had been the case with Steph Curry, big colleges had taken no interest in him. A total lack of appeal to college basketball coaches was now a good thing! Harvey was taken late in the second round of the draft with the 51st overall pick. "If Curry doesn't exist, no way he [Harvey] is drafted," said Morey.

pictures form in the minds of basketball experts which later proved a mirage. The picture in virtually every professional basketball scout's mind of Jeremy Lin, for instance. The now world-famous Chinese American shooting guard graduated from Harvard in 2010 and entered the NBA draft. "He lit up our model," said Morey. "Our model said take him with, like, the 15th pick in the draft." The objective measurement of Jeremy Lin didn't square with what the experts saw when they watched him play: a not terribly athletic Asian kid. Morey hadn't completely trusted his model — and so had chickened out and not drafted Lin. A year after the Houston Rockets failed to draft Jeremy Lin, they began to measure the speed of a player's first two steps: Jeremy Lin had the quickest first move of any player measured. He was explosive and was able to change direction far more quickly than most NBA players. "He's incredibly athletic," said Morey. "But the reality is that every fucking person, including me, thought he was unathletic. And I can't think of any reason for it other than he was Asian."

In some strange way people, at least when they were judging other people, saw what they expected to see and were slow to see what they hadn't seen before. How bad was the problem? When Jeremy Lin's coach at the New York Knicks finally put him in the game — because everyone else was injured — and allowed him to light up Madison Square Garden, the Knicks were preparing to release Jeremy Lin. Jeremy Lin had already decided that if he was released he'd simply quit basketball altogether. That's how bad the problem was: that a very good NBA player would never have been given a serious chance to play in the NBA, simply because the minds of experts had concluded he did not belong. How many other Jeremy Lins were out there?

After the Houston Rockets and everyone else in the NBA neglected to see Jeremy Lin's value in the draft (he signed after the draft as a free agent), the league shut down. A dispute between players and owners led to a lockout, and no one was allowed to work. Morey enrolled in an executive education course at Harvard Business School and took a class in behavioral economics. He'd heard of the discipline ("I'm not an idiot") but had never studied it. At the start of the first class, the professor asked him and everyone else in the class to write down the last two digits of their cell phone on a sheet of paper. Then she asked the class to write down their best estimate of the number of

African countries in the United Nations. Then she collected all the papers and showed them that the people whose cell phone numbers were higher offered systematically higher estimates of African countries in the United Nations. Then she took another example and said, "I'm going to do it again. I'm about to anchor you. Here. See if you aren't screwed up." Everyone had been warned; everyone's minds remained screwed up. Simply knowing about a bias wasn't sufficient to overcome it: The thought of that made Daryl Morey uneasy.

When the NBA returned to work he made yet another unsettling discovery. Just before the draft, the Toronto Raptors called and offered to trade their high first-round draft pick for Houston's backup point guard, Kyle Lowry. Morey talked about it with his staff, and they were on the brink of not doing the deal when one of the Rockets executives said, "You know, if we had the pick we're thinking of trading for and they offered Lowry for it, we wouldn't even consider it as a possibility." They stopped and analyzed the situation more closely: The expected value of the draft pick exceeded, by a large margin, the value they placed on the player they'd be giving up for it. The mere fact that they owned Kyle Lowry appeared to have distorted their judgment about him.* Looking back over the previous five years, they now saw that they'd systematically overvalued their own players whenever another team tried to trade for them. Especially when offered the chance to trade one of their NBA players for another team's draft picks, they'd refused deals they should have done. Why? They hadn't done it consciously.

Morey thus became aware of what behavioral economists had labeled "the endowment effect." To combat the endowment effect, he forced his scouts and his model to establish, going into the draft, the draft pick value of each of their own players.

The next season, before the trade deadline, Morey got up before his staff and listed on a whiteboard all the biases he feared might distort their judgment: the endowment effect, confirmation bias, and others. There was what people called "present bias" — the tendency, when making a decision, to undervalue the future in relation to the present. There was "hindsight bias" — which he thought of as the tendency for

* They made the trade, and then used the draft pick as the biggest chit in a deal to land a superstar, James Harden.

people to look at some outcome and assume it was predictable all along. The model was an antidote to these vagaries of human judgment, but, by 2012, the model seemed to be approaching a limit to the informational edge it would give the Rockets in valuing players. "Every year we talk about what to take out and what to put in the model," said Morey. "And every year it gets a little more depressing."

This job of running a professional basketball team had turned out to be a bit different than he had imagined, back when he was a kid. It was as if he had been assigned to take apart a fiendishly complicated alarm clock to see why it wasn't working, only to discover that an important part of the clock was inside his own mind.

Morey and his staff had obviously seen a lot of big men. But in the winter of 2015, even they were shocked by the sight of the Indian who walked into their interview room. He was dressed simply in sweatpants and a lime-green Nike T-shirt, with a pair of dog tags dangling from his neck. That neck — like his hands, his feet, his head, and even his ears — was so cartoonishly immense that you found your eyes jumping from feature to feature and wondering if that specific body part broke a Guinness book record. The Rockets once employed a seven-foot-six-inch Chinese center named Yao Ming whose size provoked these weird reactions in others. People would see him and turn and run, or burst out laughing, or weep. From head to toe the Indian was a few inches shorter than Yao Ming, but in every other way he was bigger. After seeing his measurements, and finding it hard to believe anyone could grow so much in just nineteen years, Morey had asked his staff to dig out his birth certificate. The Indian's agent had come back and said that the village in which he'd been born kept no birth records. Hearing this, Morey recalled what Dikembe Mutombo had once told him. Mutombo was a seven-foot-two-inch shot blocker who had come to the Rockets by way of Congo, with stops in between at five other NBA teams. He said that whenever some huge guy from overseas turned up claiming to be a lot younger than he looked, "You need to cut open his legs and count the rings."

The Indian's name was Satnam Singh. In all but his size he seemed young. He had the social uncertainty of an adolescent confused to find himself suddenly so far away from home. He smiled nervously and lowered himself into the chair at the head of the table.

"You doing all right?" said the Rockets interviewer.

"Yeah, I'm good good good." It wasn't a voice but a foghorn. So guttural it took a moment to work out what he'd said.

"We just want to get to know you a bit better," said the interviewer. "Tell us about your agent and why you selected him."

Satnam Singh rambled on nervously for a couple of minutes. It was unclear whether anyone in the room followed what he said. They gathered that he'd basically been taken care of since he was fourteen by people who imagined an NBA career for him.

"Tell us about where you are from and your family?" the interviewer asked.

His father worked on a farm. His mother was a cook. "I came here, I can't speak English," he said. "I could not speak to anyone. It was very hard for me. Nothing. Zero." As he struggled to relate the incredible story of his journey from his eight-hundred-person Indian village to the front office of the Houston Rockets, his eyes searched the room for approval. The executives of the Houston Rockets were ciphers.

Not unfriendly, but not giving up anything, either.

"What would you say your basketball strengths are?" asked the interviewer. "What are you best at?"

The Rockets interviewer read from a script. Singh's answers would be entered into the Rockets database, compared to the answers given by a thousand other players, and studied for patterns. They still clung to the hope that they might one day measure character, or at least get a sense how a poor kid would behave after he'd been handed millions of dollars and, usually, a seat on the bench. Would he keep working hard? Would he listen to coaches?

Morey hadn't found anyone — inside or outside basketball — who could answer those questions, though there was no end to psychologists who pretended to be able to. The Rockets had hired a bunch of them. "It's been horrible," says Morey. "A horrible experience. Every year I think there's got to be something there. Every year we find someone with a different approach. Every year it is totally pointless. And every year we try again. I'm starting to think psychologists are complete charlatans." The last psychologist who showed up claiming to be able to predict behavior had essentially used the Myers-Briggs personality test — and then tried to persuade Morey, after the fact, that he had warded off all manner of unseen problems. The way he'd gone on reminded Daryl Morey of a joke. "The guy walks around with a banana in his ear. And people are like, 'Why do you have a banana in your ear?' He says, 'To keep the alligators away! There are no alligators! See?' "

The Indian giant said his strengths were his post-up game and his midrange shooting.

"Have you broken any team rules while at IMG?" asked the interviewer.

Singh was confused. He didn't understand the question.

"No problems with the police?" Morey said helpfully.

"No fighting?" asked the interviewer.

Singh's face cleared. "Never!" he exclaimed. "Never in my life. I've never tried. If I tried, somebody would die."

The Rockets executives had been studying Singh's body. One of them finally couldn't contain himself. "Have you always been so tall?" he asked, going off script. "Or was there an age when you started to grow up faster?"

Singh explained that he was five foot nine at the age of eight and seven foot one at the age of fifteen. It ran in the family. His grandmother was six foot nine . . .

Morey stirred in his seat. He wanted to get back to questions that might lead to predictions. He asked, "What have you improved the most at — what can you do well now that maybe you didn't do as well two years ago?"

"I feel most badly on my mind. My mind."

"Sorry, I mean basketball skills. Like on the floor."

"Post game," he said. He said other things but they were unintelligible.

"Who do you think you are most like in the NBA — similar in terms of game?" asked Morey.

"Jowman and Shkinoonee," said Singh, without missing a beat.

A silence followed. Then Morey realized. "Oh, Yao Ming." Another pause. "Who was the second one?"

"Shkinoonee."

Someone made a guess: "Shaq?"

"Shaq, yes," said Singh, relieved.

"Oh, Shaquille O'Neal," said Morey, finally getting it.

"Yes, same body type and same post-up," said Singh. Most players compared themselves to someone they actually looked like. Then again, there was no NBA player who looked like Satnam Singh. If he made it, he'd be the league's first Indian.

"What do you got around your neck there?" Morey asked.

Singh grabbed his dog tags and stared down at his chest. "This is my family names," he said, fingering one. Then he took the second dog tag and simply read what it said: "I miss my coaches. I love ball. Ball is my life."

That he needed a dog tag to remind him wasn't the best sign. A lot of big guys played just because they were big. Long ago some coach or parent had yanked them onto a basketball court, and social pressure kept them there. They were less likely than small players to work hard to improve, and more likely to take your money and fade away. It wasn't that they were consciously deceitful; it was that the sort of big kid who had played basketball his entire life mainly to please others had become so practiced at telling people what they wanted to hear that he didn't know his own heart.

At length, Singh left the interview room. "Have we found evidence he has played organized basketball anywhere?" Morey asked, once he was gone. You couldn't control how you felt about the player after the interview, but you could use data to control the influence of those feelings. (Or could you?)

"They say he played at the IMG Academy in Florida."

"I hate these kinds of bets," said Morey. He'd watch Singh work out for thirty minutes, but his decision was already made. They had no data on him. Without data, there's nothing to analyze. The Indian was DeAndre Jordan all over again; he was, like most of the problems you faced in life, a puzzle, with pieces missing. The Houston Rockets would pass on him — and be shocked when the Dallas Mavericks took him in the second round of the NBA draft. Then again, you never knew.*

And that was the problem: You never *knew*. In Morey's ten years of using his statistical model with the Houston Rockets, the players he'd drafted, after accounting for the draft slot in which they'd been taken, had performed better than the players drafted by three-quarters of the other NBA teams. His approach had been sufficiently effective that other NBA teams were adopting it. He could even pinpoint the moment when he felt, for the first time, imitated. It was during the 2012 draft, when the players were

* As of this writing, it is still too early to tell.

picked in almost the exact same order the Rockets ranked them. "It's going straight down our list," said Morey. "The league was seeing things the same way."

And yet even Leslie Alexander, the only owner with both the inclination and the nerve to hire someone like him back in 2006, could grow frustrated with Daryl Morey's probabilistic view of the world. "He will want certainty from me, and I have to tell him it ain't coming," said Morey. He'd set out to be a card counter at a casino blackjack table, but he could live the analogy only up to a point. Like a card counter, he was playing a game of chance. Like a card counter, he'd tilted the odds of that game slightly in his favor. Unlike a card counter — but a lot like someone making a life decision — he was allowed to play only a few hands. He drafted a few players a year. In a few hands, anything could happen, even with the odds in his favor.

At times Morey stopped to consider the forces that had made it possible for him — a total outsider who could offer his employer only slightly better odds of success — to run a professional basketball team. He hadn't needed to get rich enough to buy one. Oddly enough, he hadn't needed to change anything about himself. The world had changed to accommodate him. Attitudes toward decision making had shifted so dramatically since he was a kid that he'd been *invited* into professional basketball to speed the change. The availability of ever-cheaper computing power and the rise of data analysis obviously had a lot to do with making the world more hospitable to the approach Daryl Morey took to it. The change in the kind of person who got rich enough to buy a professional sports franchise also had helped. "The owners often made their money from disrupting fields where most of the conventional wisdom is bullshit," said Morey. These people tended to be keenly aware of the value of even slight informational advantages, and open to the idea of using data to gain those advantages. But this raised a bigger question: Why had so much conventional wisdom been bullshit? And not just in sports but across the whole society. Why had so many industries been ripe for disruption? Why was there so much to be undone?

It was curious, when you thought about it, that such a putatively competitive market as a market for highly paid athletes could be so inefficient in the first place. It was strange that when people bothered to measure what happened on the court, they had measured the wrong things so happily for so long. It was bizarre that it was even possible for a total outsider to walk into the game with an entirely new approach to valuing basketball players and see his approach adopted by much of the industry.

At the bottom of the transformation in decision making in professional sports — but not only in professional sports — were ideas about the human mind, and how it functioned when it faced uncertain situations. These ideas had taken some time to seep into the culture, but now they were in the air we breathed. There was a new awareness of the sorts of systematic errors people might make — and so entire markets might make — if their judgments were left unchecked. There were reasons basketball experts could not see that Jeremy Lin was an NBA player, or could be blinded to the value of Marc Gasol by a single photograph of him, or would never see the next Shaquille O'Neal if he happened to be an Indian. "It was like a fish not knowing he is brcathing watcr unlcss somconc points it out," Morey said of people's awareness of their own mental processes. As it happens, someone had pointed it out.

2
THE OUTSIDER

Of Danny Kahneman's many doubts the most curious were the ones he had about his own memory. He'd delivered entire semesters of lectures straight from his head without a note. To his students he'd seemed to have memorized entire textbooks, and he wasn't shy about asking them to do it, too. And yet when he was asked about some event in his past, he'd say that he didn't trust his memory and so you shouldn't, either. Possibly this was a simple extension of what amounted to Danny's life strategy of not trusting himself. "His defining emotion is doubt," said one of his former students. "And it's very useful. Because it makes him go deeper and deeper and deeper." Or maybe he just wanted another line of defense against anyone hoping to figure him out. In any case, he kept at a great distance the forces and events that had shaped him.

He might not trust his memories, but he still had a few. For instance, he remembered the time in late 1941 or early 1942 — at any rate, a year or more after the start of the German occupation of Paris — when he was caught on the streets after curfew. The new laws required him to wear the yellow Star of David on the front of his sweater. His new badge caused him such deep shame that he took to going to school half an hour early so that the other children wouldn't see him walking into the building wearing it. After school, on the streets, he'd turn his sweater inside out.

Heading home too late one evening, he saw a German soldier approaching. "He was wearing the black uniform that I had been told to fear more than others — the one worn by specially recruited SS soldiers," he recalled, in the autobiographical statement required of him by the Nobel Committee. "As I came closer to him, trying to walk fast, I noticed that he was looking at me intently. Then he beckoned me over, picked me up, and hugged me. I was terrified that he would notice the star inside my sweater. He was speaking to me with great emotion, in German. When he put me down, he opened his wallet, showed me a picture of a boy, and gave me some money. I went home more certain than ever that my mother was right: people were endlessly complicated and interesting."

He also remembered the sight of his father after he'd been taken away in a big sweep in November 1941. Thousands of Jews were rounded up and sent to camps. Danny had complicated feelings about his mother. His father he'd simply loved. "My father was radiant; he had enormous charm." He was jailed in the makeshift prison in Drancy, outside of Paris. In Drancy, public housing designed for seven hundred people was used to imprison as many as seven thousand Jews at a time. "I have this memory of going with my mother to see this prison," Danny recalled. "And I remember it was sort of pink-orange. There were people, but you couldn't see the faces. You could hear women and children. And I remember the prison guard. He said, 'It's hard in there. They are eating peels.' " For most Jews, Drancy was just a stop on the way to a concentration camp: Upon arrival, many of the children were separated from their mothers and put on trains to be gassed at Auschwitz.

Danny's father was released after six weeks, thanks to his association with Eugène Schueller. Schueller was the founder and head of the giant French cosmetics company L'Oréal, where Danny's father worked as a chemist. Long after the war Schueller would be exposed as one of the architects of an organization to help the Nazis find and kill French Jews. Somehow he carved out in his mind a special exemption for his star chemist; he persuaded the Germans that Danny's father was "central to the war effort," and he was sent back to Paris. Danny recalled that day vividly. "We knew he was coming back so we went shopping. When we came back we rang the bell and he opened the door. And he was wearing his best suit. He weighed forty-five kilos [ninety-nine pounds]. He was skin and bones. And he hadn't eaten. That is the thing that impressed me. He waited for us to eat."

Seeing that even Schueller couldn't keep them safe in Paris, Danny's father took his family and fled. By 1942 the borders were closed, and there was no clear path to safety. Danny, his older sister, Ruth, and his parents, Ephraim and Rachel, made a run for the south, which the Vichy regime still nominally governed. Along the way there were close calls and complications. They hid in barns: Danny remembered those, along with the phony identity cards his father had somehow secured in Paris that contained a misspelling. Danny and his sister and mother were called "Cadet" while his father had been given the name "Godet." To avoid detection Danny had been required to call his father "Uncle." He also needed to do the speaking for his mother, as her first language was Yiddish, and she still spoke French with an accent. His mother on mute was a rare sight. She always had a great deal to say. She blamed her husband for their circumstances. They'd stayed in Paris only because he had allowed himself to be misguided by his memory of the Great War. The Germans hadn't gotten to Paris then, he'd said, so they surely wouldn't get to Paris now. She hadn't agreed. "I do remember that my mother saw the horrors coming long before he did — she was the pessimist and the worrier, he was sunny and optimistic." Danny sensed already that he was very like his mother and not at all like his father. His feelings about himself were complicated.

The approaching winter of 1942 found them in a coastal town called Juan-les-Pins, in a state of dread. They now had their own house, courtesy of the Nazi collaborator, with a chemistry lab in it, so that Danny's father could continue to work. To blend into their new society, his parents sent Danny to school, with a warning to be careful not to say too much or seem too clever. "They were afraid I would be identified as Jewish." For as long as he could remember he had thought of himself as precocious and bookish. His body he felt little connection to. He was so bad at sports he'd one day be referred to by classmates as The Living Corpse. A gym teacher would prevent him from being given academic honors on the grounds that "there are limits to everything." His mind, however, was limber and muscular. From the moment he thought of what he might be when he grew up, he simply assumed he would be an intellectual. That was his image of himself: a brain without a body. He now had a new one: a rabbit in a rabbit hunt. The goal simply was to survive.

On November 10, 1942, the Germans moved into the south of France. German soldiers in black uniforms now pulled men off buses and stripped them to see if they were circumcised. "Anyone who was caught was dead," recalled Danny. His father firmly did not believe in God: His loss of faith had led him, as a young man, to leave Lithuania, and the illustrious line of rabbis from which he descended, for Paris. Danny wasn't ready to abandon the idea that the universe had some unseen caring force in it. "I was sleeping under the same mosquito net as my parents," he said. "They were in a big bed. I was in a small bed. I was nine. And I would pray to God. And the prayer was: I know you are very busy and that this is a tough time and all that. I don't want to ask for much but I want to ask for one more day."

Again they fled for their lives, this time up the Côte d'Azur to Cagnes-sur-Mer, to a place owned by a colonel in the old French army. For the next few months Danny was confined to quarters. He passed the time with books. He read and reread *Around the World in Eighty Days* and fell in love with all things English and, especially, with Phileas Fogg. The French colonel had left behind a long shelf filled with accounts of the trench warfare at Verdun, and Danny read all of those, too — and became something of an expert on the subject. His father still worked in the house down the coast with the chemistry lab in it, traveling by bus to see his family on weekends. On Fridays Danny sat with his mother in the garden and watched her darn socks and waited for his father to arrive. "We lived on the hill and we could see the bus station. We never knew if he would come. I have hated waiting ever since."

With help from the Vichy government and private bounty hunters, the Germans became more efficient at hunting Jews. Danny's father suffered from diabetes, but it was now more dangerous for him to seek treatment for it than to live with it untreated. Once again they ran. First to hotels and then, finally, to the chicken coop. The chicken coop was behind a country bar in a small village outside Limoges. Here there were no German soldiers, only the Milice — the paramilitary force collaborating with the Germans to help them round up Jews and exterminate the French Resistance. How his father had found the place Danny didn't know, but L'Oréal's founder must have been involved, as the company continued to send packages of food. They erected a partition in the middle of the room so Danny's sister might have some privacy, but the coop wasn't really meant for anyone to live in. In winter it grew so cold the door froze shut. His sister tried to sleep on the stove and ended up with burn marks on her robe.

To pass as Christians, Danny's mother and sister went to church on Sundays. Danny, now ten years old, returned to school, on the theory that he was less conspicuous there than hiding inside the chicken coop. The students at this new country school were even less able than the ones in Juan-les-Pins. The teacher was kind but forgettable. The only lesson Danny recalled was the one about the facts of life. He found the details so preposterous that he was sure the teacher had been mistaken. "I said, 'That is absolutely impossible!' I asked my mother about it. She said it was so." Still, he didn't really believe it until one night when he was in bed, with his mother sleeping beside him. Waking up and needing to use the outhouse, he climbed over her. She awakened to find her son on top of her. "And my mother is terrified. And I think, 'It must be true after all!' "

Even as a child he had an almost theoretical interest in other people — why they thought what they thought, why they behaved as they did. His direct experience of them was limited. He attended school but avoided social contact with his teachers and classmates. He had no friends. Even acquaintances were life-threatening. On the other hand, he witnessed, from a certain distance, a lot of interesting behavior. Both his teacher and the owners of the local bar, he had to believe, couldn't help but know that he was Jewish. Why else would this precocious ten-year-old city boy land in a schoolroom filled with country bumpkins? Why else would this clearly well-heeled family of four have piled into a chicken coop? Yet they gave him no sign they were anything but oblivious. His teacher gave him high marks and even invited Danny to his home, and Madame Andrieux, who owned the bar, asked him to help out, gave him tips (for which he had no use), and even tried to talk his mother into opening a brothel with her. A lot of other people quite obviously failed to see them for what they were. Danny remembered in particular the young French Nazi, a member of the Milice, who courted, without success, Danny's sister. She was now nineteen, with movie star looks. (After the war, she took great pleasure in letting the Nazi know that he had fallen in love with a Jew.)

On the night of April 27, 1944 — that date Danny remembered clearly — his father took him for a walk. He now had dark spots inside his mouth. Forty-nine years old, he looked much older. "He told me I might have to become responsible," recalled Danny. "He told me to think of myself as the man of the family. He told me how to try to keep things under control with my mother — that I was sort of the sane one in the family. I had a book of poems I'd written. And I gave them to him. And he died that night." Of his father's death Danny had little memory except that his mother had made him spend the night with Monsieur and Madame Andrieux. There was another Jew hiding in their village. His mother had found him and he had helped remove his father's body before Danny returned. She gave him a Jewish burial but didn't invite Danny to attend, probably because it was so dangerous. "I was really angry about him dying," said Danny. "He had been good. But he had not been strong."

The Allies invaded Normandy six weeks later. Danny never saw any soldiers. No American tanks rolled through his village with GIs on top tossing candy to children. One day he woke up and there was a feeling of joy in the air and the Milice were being marched off to be shot or jailed, and a lot of women were walking around with shaved heads — punishment for having slept with a German. By December the Germans had been driven out of France, and Danny and his mother were free to travel to Paris to see what remained of their home and chattels. Danny kept a notebook, which he had titled "What I Write of What I Think." ("I must have been intolerable.") In Paris he read, in one of his sister's schoolbooks, an essay by Pascal that inspired him to write in his notebook an essay of his own. The Germans were then launching their final counterattack to retake France, and Danny and his mother lived with the fear that they would break through: Danny wrote an essay that attempted to explain man's need for religion. He began with a quote from Pascal, *Faith is God made sensible to the heart,* then added, "How true!" He followed this up with his own original line: "Cathedrals and organs are artificial ways of generating the same feeling." He no longer thought of God as an entity to which he might pray. Later, when he looked back on his life, he remembered his childhood pretensions and was both proud of and embarrassed by them. His precocious essay writing, he thought, was "deeply linked in my mind with knowing that I was a Jew, with just a mind and no useful body, and that I would never fit in with other boys."

In Paris, in their old prewar apartment, Danny and his mother found only two battered green chairs. Still, they stayed. For the first time in five years Danny attended school without having to disguise who he was. For years he carried a fond memory of the friendship he struck up there with a pair of tall, handsome Russian aristocrats. The memory was so insistent, perhaps, because he had gone so long without friends. Much later in life, he tested his memory by tracking down the aristocratic Russian brothers and sending them a note. One brother had become an architect, the other a doctor. The brothers wrote back to say that they remembered him, and sent him a picture of them all together. Danny wasn't in the picture: They must have been thinking he was somebody else. His lone friendship was imagined, not real.

The Kahnemans no longer felt welcome in Europe and left in 1946. Danny's father's extended family had remained in Lithuania and, along with the six thousand or so other Jews in their city, had been slaughtered. Only Danny's uncle, a rabbi, who happened to be out of the country when the Germans rolled in, had been spared. He, like Danny's mother's family, now lived in Palestine — and so to Palestine they moved. Their arrival was sufficiently momentous that someone filmed it (the film was lost), but all Danny would later say he remembered of it was the glass of milk his uncle brought him. "I still remember how white it was," he said. "It was my first glass of milk in five years."

Danny and his mother and sister moved in with his mother's family in Jerusalem. There, a year later, at the age of thirteen, Danny made his final decision about God. "I still remember where I was — the street in Jerusalem. I remember thinking that I could imagine there was a God, but not one who cared whether or not I masturbate. I reached the conclusion that there was no God. That was the end of my religious life."

And that's pretty much what Danny Kahneman remembered, or chose to remember, when asked about his childhood. From the age of seven he had been told to trust no one, and he'd obliged. His survival had depended on keeping himself apart, and preventing others from seeing him for what he was. He was destined to become one of the world's most influential psychologists, and a spectacularly original connoisseur of human error. His work would explore, among other things, the role of memory in human judgment. How, for instance, the French army's memory of Germany's military strategy in the last war might lead them to misjudge that strategy in a new war. How a man's memory of German behavior in one war might lead him to misjudge Germans' intentions during the next. Or how the memory of a little boy back in Germany might prevent a member of Hitler's SS, trained to spot Jews, from seeing that the little boy he has picked up in his arms from the streets of Paris is a Jew.

His own memories he didn't find all that relevant, however. For the rest of his life he insisted that his past had little effect on his view of the world or, ultimately, the world's view of him. "People say your childhood has a big influence on who you become," he'd say, when pressed. "I'm not at all sure that's true." Even to those he came to regard as his friends he never mentioned his Holocaust experience. Really, it wasn't until after he won the Nobel Prize and journalists started to badger him for the details of his life that he began to offer them up. His oldest friends would learn what had happened to him from the newspaper.

The Kahnemans had arrived in Jerusalem just in time for another war. In the fall of 1947 the problem of Palestine passed from Britain to the United Nations, which, on November 29, passed a resolution that formally divided the land into two states. The new Jewish state would be roughly the size of Connecticut and the Arab state just a bit smaller than that. Jerusalem, and its holy sites, belonged to neither. Anyone living in Jerusalem would become a "citizen" of Jerusalem; in practice, there was an Arab Jerusalem and a Jewish Jerusalem, and the residents of each continued to do their best to kill each other. The apartment into which Danny moved with his mother was near the unofficial border: A bullet passed through Danny's bedroom. The leader of his scout troop was killed.

And yet, Danny said, life didn't feel particularly dangerous. "It was so completely different. Because you are fighting. That is why it is better. I *hated* the status of being a Jew in Europe. I didn't want to be hunted. I didn't want to be a rabbit." Late one night in January 1948 he saw, with a palpable thrill, his first Jewish soldiers: thirty-eight young fighters gathered in the basement of his building. Arab fighters had blockaded a cluster of Jewish settlements in the south of the tiny country. The thirty-eight Jewish soldiers marched off from Danny's basement to rescue the settlers. Along the way, three turned back one who had sprained an ankle, and two others to help him walk home — and so the group would become known for all time as "The 35." They'd intended to march under cover of darkness, but the sun rose to find them still marching. They met an Arab shepherd and decided to let him go — at least that is the story that Danny heard. The shepherd informed the Arab fighters, who ambushed and killed all thirty-five young men and then mutilated their bodies. Danny wondered at their disastrous decision. "Do you know why they were killed?" he said. "They were killed because they could not bring themselves to shoot a shepherd."

A few months later, a convoy of doctors and nurses under the Red Cross banner drove the narrow road from the Jewish city to Mt. Scopus, the site of Hebrew University and the hospital attached to it. Mt. Scopus lay behind Arab lines, a Jewish island in a sea of Arabs. The only way in was through a mile-and-a-half-long narrow road over which the British guaranteed safe passage. Most of the time the trip was uneventful, but on this day a bomb exploded and stopped the lead vehicle, a Ford truck. Arab machine-gun fire raked the buses and ambulances that followed. A few of the cars in the convoy were able to turn and speed off, but the buses, which carried passengers, were trapped. When the shooting stopped, seventy-eight people were dead, their bodies so badly burned that they were buried in a mass grave. Among them was Enzo Bonaventura, a psychologist imported from Italy nine years earlier by Hebrew University to build a department of psychology. His plans for a psychology department died with him.

Whatever threat Danny felt to his existence he declined to acknowledge. "It looked very implausible — that we would defeat five Arab nations — but somehow we were not worried. There really was no sense of impending doom that I could pick up. People were killed and so on. But, for me, after World War II, it was a picnic." His mother evidently did not agree, as she took her fourteen-year-old son and fled Jerusalem for Tel Aviv.

On May 14, 1948, Israel declared itself a sovereign state, and the British soldiers left the next day. The armies of Jordan, Syria, and Egypt attacked, along with some troops from Iraq and Lebanon. For many months Jerusalem was under siege, and life in Tel Aviv was far from normal. The minaret on the beach beside what is now the Intercontinental Hotel became an Arab sniper nest: The sniper could, and did, shoot at Jewish children on their way to and from school. "There were bullets flying everywhere," recalled Shimon Shamir, who was fourteen years old and living in Tel Aviv when the war broke out, and would grow up to become the only person ever to serve as

Israel's ambassador to both Egypt and Jordan.

Shamir was Danny's first real friend. "The other kids in class felt there was some distance between them and him," said Shamir. "He wasn't looking for groups. He was very selective. He didn't need more than one friend." Danny spoke no Hebrew when he arrived in Israel the year before, but by the time he arrived at school in Tel Aviv he spoke it fluently, and spoke English better than anyone else in the class. "He was considered brilliant," says Shamir. "I used to tease him: 'You are going to be famous.' And he would feel very uncomfortable about it. I hope I am not reading history back, but I think there was a feeling that he would go a long way."

It was clear to all that Danny wasn't like the other boys. He wasn't trying to be unusual; he just was. "He was the only one in our class who tried to develop a proper English accent," said Shamir. "We all found that very funny. He was different in many ways. To some extent he was an outsider. And it was because of his personality, not because he was a refugee." Even at the age of fourteen Danny was less a boy than an intellectual trapped in a boy's body. "He was always absorbed in some problem or question," said Shamir. "I remember one day he showed me a long essay he wrote for himself — which was strange, because writing essays was a burden which you only did for school, on the subject the teacher assigned. The whole idea of writing a very long essay on a subject that had nothing to do with the curriculum just because the subject interested him: That impressed me very much. He compared the personality of an English gentleman with that of a Greek aristocrat at the time of Herakles." Shamir felt that Danny was searching books and his own mind for a direction most children get from the people around them. "I think he was looking for an ideal," he said. "A role model."

The war of independence lasted for ten months. A Jewish state that was the size of Connecticut before the war wound up a bit bigger than New Jersey. One percent of the Israeli population had been killed (the equivalent of ninety thousand dead in New Jersey). Ten thousand Arabs had died, and three-quarters of a million Palestinians were displaced. After the war, Danny's mother moved them back to Jerusalem. There Danny made his second close friend, a boy of English descent named Ariel Ginsberg.

Tel Aviv was poor, but Jerusalem was even

poorer. Basically no one owned a camera, or a phone, or even a doorbell. If you wanted to see a friend you had to walk to his house and knock on the door or whistle. Danny would walk to Ariel's house, whistle, and Ariel would come down and they'd head to the YMCA to swim and play Ping-Pong without uttering a word. Danny thought that was just perfect: Ginsberg reminded him of Phileas Fogg. "Danny was different," says Ginsberg. "He felt apart and he kept himself apart — up to a point. I was his only friend."

In just a few years after the war of independence, the Jewish population of what was now called Israel doubled, from 600,000 to 1.2 million. There can have been no time or place on earth where it was easier and more strongly encouraged for a Jewish person newly arrived in a country to assimilate into the local population. And yet, in spirit, Danny did not assimilate. The people to whom he gravitated were all native-born Israelis rather than fellow immigrants. But he himself did not seem Israeli. Like many Israeli boys and girls, he joined the scouts — then quit when he and Ariel decided the group was not for them. Although he'd learned Hebrew with incredible speed, he and his mother spoke French at home, often in angry tones. "It was not a happy home," says Ginsberg. "His mother was a bitter woman. His sister got out of there as fast as she could." Danny didn't accept Israel's offer of a new pre-packaged identity. He accepted its offer of a place to create his own.

What that identity would be was hard to pin down, because Danny himself was so hard to pin down: He didn't seem to wish to settle anywhere in particular. What attachments he formed felt loose and provisional. Ruth Ginsberg, who was then dating and would soon marry Danny's close friend, said, "Danny decided very early on that he would not take responsibility. I had the feeling that there was a need within him to always rationalize his unrootedness. A person who does not need roots. To have this view of life as a series of coincidences — *it happened this way but it could just as well have happened some other way.* You make the best of it within these godless conditions."

Danny's lack of need for a place or a group to belong to was especially glaring in a land of people hungry for a place and a people. "I came in 1948 and I wanted to be like they are," recalls Yeshu Kolodny, a professor of geology at Hebrew University,

Danny's age, whose extended family also had been wiped out in the Holocaust. "Meaning I wanted to wear sandals and shorts rolled up and learn the name of every goddamn wadi [valley] or mountain — and mainly I wanted to lose my Russian accent. I was a little bit ashamed of my story. I came to worship the heroes of my people. Danny didn't feel that way. He looked down on this place."

Danny was a refugee in the way that, say, Vladimir Nabokov was a refugee. A refugee who kept his distance. A refugee with *airs.* And a sharp eye for the locals. At the age of fifteen he took a vocational test that identified him as a psychologist. It didn't surprise him.* He'd always sensed that he would be

* Decades later, when Danny Kahneman was in his forties, he sat in for a day on a class at the University of California, Berkeley, taught by a psychologist named Eleanor Rosch. On that day, Rosch put a group of first-year graduate students through an exercise. She passed around a hat stuffed with slips of paper, on each slip a different occupation: zookeeper, airline pilot, carpenter, thief. The students were told to pick an occupation and then say what, if anything, popped to mind that foreshadowed their fate. *Of course I wound up a zookeeper; as a kid I loved to cage our*

some sort of professor, and the questions he had about human beings were more interesting to him than any others. "My interest in psychology was as a way to do philosophy," he said. "To understand the world by understanding why people, especially me, see it as they do. By then the question of whether God exists left me cold. But the question of why people believe God exists I found really fascinating. I was not really interested in right and wrong. But I was very interested in indignation. Now *that*'s a psychologist!"

Most Israelis, upon finishing high school, were conscripted into the military. Identified as intellectually gifted, Danny was allowed to proceed directly to university to

---

cat. The exercise was meant to illustrate the powerful instinct people have for finding causes for any effect, and also for creating narratives. "The whole group opens their papers at the same time," recalled Rosch, "and within seconds someone laughs, and the laughter becomes general. And, yes, to their surprise, things have popped into their minds. Danny was the lone exception. " 'Nope,' he said," according to Rosch. " 'I could only have been two things. A psychologist or a rabbi.' "

pursue a degree in psychology. How to do this was not obvious, as the country's only college campus lay behind Arab lines, and its plans for a psychology department had been killed in an Arab ambush. And so, on a morning in the fall of 1951, the seventeen-year-old Danny Kahneman sat in math class, held in a Jerusalem monastery that served as one of several temporary homes for Hebrew University. Even here, Danny seemed out of place. Most of the students had just come from serving three years in the army, and a lot of them had seen combat. Danny was younger, and dressed in a jacket and tie, which struck the other students as preposterous.

For the next three years Danny essentially taught himself great swaths of his chosen field, as his teachers could not. "My statistics teacher I loved," Danny recalled, "but she didn't know statistics from beans. I taught myself statistics, from a book." His professors were less an assemblage of specialists than a collection of characters, most of them European refugees, who happened to be willing to live in Israel. "Basically it was organized around charismatic teachers, people who had biographies, not just curriculum vitaes," recalled Avishai Margalit, who would go from Hebrew University to become a philosophy professor at Stanford, among other places. "They had lived big lives."

The most vivid was Yeshayahu Leibowitz — whom Danny adored. Leibowitz had come to Palestine from Germany via Switzerland in the 1930s, with advanced degrees in medicine, chemistry, the philosophy of science and — it was rumored — a few other fields as well. Yet he'd tried and failed to get his driver's license seven times. "You'd see him walking the streets," recalled one former Leibowitz student, Maya Bar-Hillel. "His pants pulled up to his neck, he had these hunched shoulders and a Jay Leno chin. He'd be talking to himself and making these rhetorical gestures. But his mind attracted youth from all over the country." Whatever Leibowitz happened to be teaching — and there seemed no subject he could not teach — he never failed to put on a show. "The course I took from him was called biochemistry, but it was basically about life," recalled another student. "A large part of thc class was devoted to explaining how stupid Ben-Gurion was." He was referring to David Ben-Gurion, Israel's first prime minister. One of Leibowitz's favorite stories was about a donkey placed equidistant from two bundles of hay.

In the story the donkey can't decide which bundle of hay is closer to him, and so dies of hunger. "Leibowitz would then say that no donkey would do this; a donkey would just go at random to one or the other and eat. It's only when decisions are made by people that they get more complicated. And then he said, 'What happens to a country when a donkey makes the decisions that people are supposed to make you can read every day in the paper.' His class was always full."

What Danny recalled of Leibowitz was typically peculiar: not so much what the man had said but the sound made by the chalk hitting the board when he wanted to make a point. It was like a gunshot.

Even at that young age, and in those circumstances, it was possible to detect a drift in Danny's mind, by the currents it resisted. Freud was in the air but Danny didn't want anyone lying on his couch, and he really didn't want to lie upon anyone else's. He'd decided to attach no particular importance to his own childhood experience, or even his memories: Why should he care about other people's? By the early 1950s, some large number of the psychologists who insisted that the discipline be subject to the standards of science had given up the ambition to study the inner workings of the human mind. If you can't observe what is happening in the mind, how can you even pretend to make a study of it? What was deemed worthy of scientific attention — and what could be studied scientifically — was how living creatures behaved.

The dominant school of thought was called behaviorism. Its king, B. F. Skinner, had gotten his start during the Second World War, after the U.S. Air Force hired him to train pigeons to guide bombs. Skinner taught his pigeons to peck in the right spot on an aerial map of the target, by rewarding them with food each time they did it. (They did this with less enthusiasm when anti-aircraft fire was exploding around them, and so were never used in combat.) Skinner's success with the pigeons was the start of a spectacularly influential career underpinned by the idea that all animal behavior was driven not by thoughts and feelings but by external rewards and punishments. He locked rats inside what he called "operant conditioning chambers" (they soon became known as "Skinner boxes") and taught them to pull levers and push buttons. He taught pigeons to dance and play Ping-Pong and bang out "Take Me Out to the Ball Game" on a piano.

The behaviorists presumed that whatever they discovered about rats and pigeons applied to people — on whom, for various reasons, it was simply less practical to conduct experiments. "To the reader who is anxious to advance to the human subject a word of caution is in order," Skinner wrote, in an essay called "How to Teach Animals." "We must embark upon a program in which we sometimes apply relevant reinforcement and sometimes withhold it. In doing this, we are quite likely [in humans] to generate emotional effects. Unfortunately the science of behavior is not yet as successful in controlling emotion as it is in shaping behavior." The allure of behaviorism was that the science appeared clean: the stimuli could be observed, the responses could be recorded. It seemed "objective." It didn't rely on anyone telling anyone else what he thought or felt. All the important stuff was observable and measurable. There was a joke that captured the antiseptic spirit of behaviorism that Skinner himself liked to tell: A couple makes love. Afterward, one of them turns to the other and says, "It was good for you. How was it for me?"

All the leading behaviorists were WASPs — a fact that didn't go unnoticed by young people entering psychology in the 1950s.

Looking back, a casual observer of the field at that time couldn't help but wonder if there shouldn't be two entirely unrelated disciplines: "WASP Psychology" and "Jewish Psychology." The WASPs marched around in white lab coats carrying clipboards and thinking up new ways to torture rats and all the while avoided the great wet mess of human experience. The Jews embraced the mess — even the Jews who disdained Freud's methods and longed for "objectivity" and wished to search for the kinds of truth that might be tested according to the rules of science.

Danny, for his part, longed for objectivity. The school of psychological thought that most charmed him was Gestalt* psychology. Led by German Jews — its origins were in early twentieth-century Berlin — it sought to explore, scientifically, the mysteries of the human mind. The Gestalt psychologists had made careers uncovering interesting phenomena and demonstrating them with great flair: a light appeared brighter when it emerged from total darkness; the

* The word is German and means "shape" or "form" but, in a manner the Gestalt psychologists would enjoy, has itself tended to change shape, depending on the context in which it is used.

color gray looked green when it was surrounded by violet and yellow if surrounded by blue; if you said to a person, "Don't step on that banana eel!," he'd be sure that you had said not "eel" but "peel." The Gestalists showed that there was no obvious relationship between any external stimulus and the sensation it created in people, as the mind intervened in many curious ways. Danny was especially struck by the way that the Gestalt psychologists, in their writings, put their readers through an experience, so that they might feel for themselves the mysterious inner workings of their own minds:

> If on a clear night we look up at the sky, some stars are immediately seen as belonging together, and as detached from their environment. The constellation Cassiopeia is an example, the Dipper is another. For ages people have seen the same groups as units, and at the present time children need no instruction in order to perceive the same units. Similarly, in figure 1 the reader has before him two groups of patches.

Figure 1. Adapted from Wolfgang Köhler, Gestalt Psychology *(1947; repr., New York: Liveright, 1992), 142.*

Why not merely six patches? Or two other groups? Or three groups of two members each? When looking casually at this pattern everyone beholds the two groups of three patches each.

The central question posed by Gestalt psychologists was the question the behaviorists had elected to ignore: How does the brain create meaning? How does it turn the fragments collected by the senses into a coherent picture of reality? Why does that picture so often seem to be imposed by the mind upon the world around it, rather than by the world upon the mind? How does a person turn the shards of memory into a coherent life story? Why does a person's understanding of what he sees change with the context in which he sees it? Why — to speak a bit loosely — when a regime bent on the destruction of the Jews rises to power in Europe, do some Jews see it for what it is, and flee, and others stay to be slaughtered? These questions, or ones like them, had led Danny into psychology. They weren't the sort to be answered by even the most gifted rat. Their answers, if they existed, could be found only in the human mind.

Later in his life Danny would say that he thought of science as a conversation. If so, psychology was a noisy dinner party during which the guests talked past one another and changed the subject with bewildering frequency. The Gestalt psychologists and the behaviorists and the psychoanalysts might all be jammed into the same building with a plaque on the front that said Department of Psychology, but they didn't waste a lot of time listening to one another. Psychology wasn't like physics, or even economics. It lacked a single persuasive theory to organize itself around, or even an agreed-upon set of rules for discussion. Its leading figures could, and did, say of the work of other psychologists, *Basically, what you are doing and saying is total bullshit,* without any discernible effect on the behavior of those psychologists.

Part of the problem was the wild diversity

of the people who wanted to be psychologists — a rattle-bag of characters with motives that ranged from the urge to rationalize their own unhappiness, to a conviction that they had deep insights into human nature but lacked the literary power to write a decent novel, to a need for a market for their math skills after they'd been found inadequate by the physics department, to a simple desire to help people in pain. The other issue was the grandma's attic quality of the field: Psychology was a place all sorts of unrelated and seemingly unsolvable problems simply got tossed. "It is possible to find two competent and highly productive academic psychologists who, if they had lunch together, would be forced to discuss the Twins' chances for the pennant or Ronald the Red Killer's showmanship talents, because they would have negligible overlap in their knowledge and interests in psychology," the University of Minnesota psychologist Paul Meehl wrote in a famous 1986 essay, "Psychology: Does Our Heterogeneous Subject Matter Have Any Unity?" "One can inquire as to why this is, whether anything can be done about it, or — a question that should be asked first — does it really matter anyway? Why *should* a behavior geneticist studying the transmission of schizophrenia be able to converse with an expert on the electrochemical processes in the retina of the walleyed pike?"

Aptitude tests revealed Danny to be equally suited for the humanities and science, but he only wanted to do science. He also wanted to study people. Beyond that, it soon became clear, he didn't know what he wanted to do. In his second year at Hebrew University, he listened to a talk by a visiting German neurosurgeon who claimed that damage to the brain caused people to lose the capacity for abstract thought. The claim turned out to be false, but Danny was so taken by it in that moment that he decided to chuck psychology to pursue a medical degree — so that he'd be allowed to poke around the human brain and see what other effects he might generate. A professor eventually persuaded him that it was insane to go through the misery of acquiring a medical degree unless he actually wanted to be a doctor. But it was the start of a pattern: seizing on some idea or ambition with great enthusiasm only to abandon it in disappointment. "I've always felt ideas were a dime a dozen," he said. "If you had one that didn't work out, you should not fight too hard to save it, just go find another."

In an ordinary society it is unlikely that

anyone would ever have discovered the fantastic practical usefulness of Danny Kahneman. Israel wasn't a normal society. Graduating from Hebrew University — which somehow bestowed upon him a degree in psychology — Danny was required to serve in the Israeli army. Gentle, detached, disorganized, conflict-avoiding, and physically inept: Danny wasn't anyone's idea of a soldier. Only twice did he come close to having to fight, and both instances remained, to him, vividly memorable. The first time came when the platoon that he and several others commanded was ordered to attack an Arab village. Danny's platoon was meant to circle around the village and ambush any Arab forces. The year before, after an Israeli army unit had massacred Arab women and children, Danny and his friend Shimon Shamir had discussed what they would do if they were ordered to kill Arab civilians. They'd decided that they would refuse the order. Here was the closest Danny would come to being given that order. "We were not supposed to go into the village," he said. "The other officers were given their orders. And I listened — and they were never told to kill civilians. But they were never told how not to kill civilians. And I couldn't ask the question —

because it wasn't my mission." In the event, his own mission was aborted and his unit withdrawn before it came anywhere near to shooting at anyone — and only later did he learn why. The other platoons had walked into an ambush. The Jordanian army had been waiting for them. Had he not withdrawn, "We would have been butchered."

The other time, he was sent one night to lay ambushes for the Jordanian army. He had three squads in his platoon. He led each of the first two squads to their ambush sites and left subordinates in charge of them. The third, on the Jordanian border, he led himself. To find the border, his commanding officer (a poet named Haim Gouri) told him, he should walk until he reached a sign: *Frontier. Stop.* In the dark, Danny missed the sign. As the sun rose, what he saw instead was an enemy soldier, on a hill, with his back to him: He'd invaded Jordan. ("I nearly started a war.") The stretch of land beneath the hill in front of them, he saw, was ideally suited for Jordanian snipers looking to pick off Israeli soldiers. Danny turned to sneak his patrol back into Israel, but then he noticed that one of his men was missing his pack. Imagining the dressing-down he'd receive for leaving a pack in Jordan, he and his men crept around the fringes of the kill zone. "It was incredibly dangerous. I knew how stupid it was. But we would stay until we found it. Because I could hear the first question, 'How could you leave that pack?' That has stayed with me: the idiocy of it." They found the pack, then left. Upon his return, his superiors admonished him, but not about the backpack. "They said, 'Why didn't you shoot?' "

The army jolted him out of his usual self-assigned role of detached observer. His year as a platoon commander, Danny said later, "removed the remaining traces of the pervasive sense of vulnerability and physical weakness and incompetence which I had had in France." But he wasn't born to shoot at people. He wasn't really suited to army life, either, but the army forced him to be suitable. They assigned him to the psychology unit. The chief feature of the Israeli army's psychology unit in 1954 was that it had no psychologists. Upon joining it, Danny found that his new boss — the Israeli army's chief of psychological research — was a chemist. So Danny, a twenty-year-old refugee from Europe who had spent a meaningful amount of his life in hiding, found himself the Israel Defense Forces' expert on psychological matters. "He was thin, ugly, and very clever," recalls Tammy

Viz, who served with Danny in the psychology unit. "I was nineteen and he was twenty-one, and I think he flirted with me and I was so dumb I didn't know it. He was not a normal guy. But people liked him." They also needed him — though how much they surely did not immediately appreciate.

The new nation faced a serious problem: how to organize a madly diverse population into a fighting force. In 1948, David Ben-Gurion had declared Israel open to any Jew who wished to immigrate. Over the next five years, the state accepted more than 730,000 immigrants from different cultures, speaking different languages. Many of the young men entering the new Israel Defense Forces already had endured unspeakable horrors — everywhere you turned, you found people with numbers tattooed on their arms. Mothers stumbled unexpectedly upon their own children, who they thought had been murdered by the Germans, on the streets of Israeli cities. No one was encouraged to speak about what he'd experienced in war. "People who had post-traumatic stress disorder were considered weaklings," as one Israeli psychologist put it. Part of the job of being an Israeli Jew was to at least pretend to forget the unforgettable.

Israel was still less a nation than a fort,

and yet its army was in a state of barely controlled chaos. The soldiers were poorly trained, the units poorly coordinated. The head of the tank division didn't even speak the same language as most of his men. In the early 1950s there was no formal war between Arabs and Jews, but the senseless metronomic violence exposed vulnerabilities in the Israeli military. The soldiers tended to cut and run at the first sign of trouble, for instance; and the officers tended to lead from behind. The infantry staged a succession of failed night raids on Arab outposts, during which Israeli troops got lost in the dark and never reached their targets. In one case, after a unit sent out to stage an attack spent the night wandering around in circles, the platoon commander had simply shot himself. When they managed to engage the enemy, the results were often disastrous. In October 1953, an Israeli unit that may or may not have been given instructions not to harm civilians had raided a Jordanian village and killed sixty-nine people, half of them women and children.

Since the First World War, the job of assessing and sorting young conscripts into armies had fallen to psychologists, mainly because some ambitious psychologists had talked the U.S. Army into giving them the job. Still, if you need to sort tens of thousands of young men quickly into an efficient fighting force, it's not immediately obvious that you also need a psychologist, and even less obvious when the only psychologist at hand is a twenty-one-year-old graduate of a two-year program who has more or less taught himself. Danny himself was surprised they asked him to do it, and did not feel equipped for the job. And he'd already seen the difficulty of trying to figure out which person was suited to which job when his superiors had asked him to evaluate candidates for officer training school.

The young men applying to become officers had been given a weirdly artificial task: to move themselves from one side of a wall to the other without touching the wall, using only a long log that was not permitted to touch either the wall or the ground. "We noted who took charge, who tried to lead and was rebuffed, how cooperative each soldier was in contributing to the group effort," Danny wrote. "We saw who seemed to be stubborn, submissive, arrogant, patient, hot-tempered, persistent, or a quitter. We saw competitive spite when someone whose idea had been rejected by the group sabotaged its efforts. And we saw reactions to crisis. . . . Under the stress of the event,

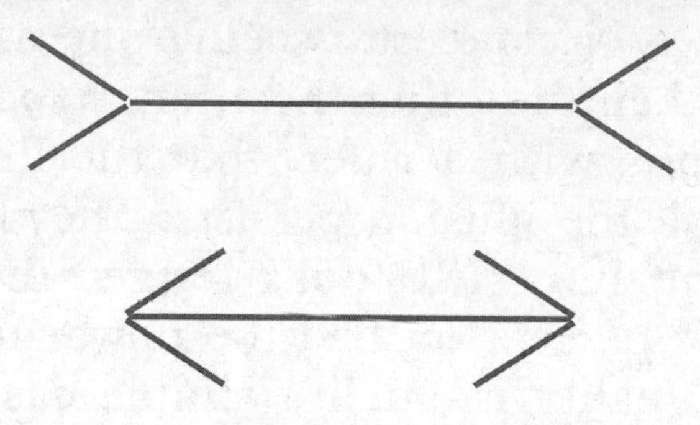

Figure 2. Müller-Lyer optical illusion.

we felt, each man's true nature was revealed. The impression we had of each candidate's character was as direct and compelling as the color of the sky."

He had had no trouble identifying which men would make good officers and which would not. "We were quite willing to declare, 'This one will never make it,' 'That fellow is rather mediocre,' or 'He will be a star.' " The problem came when he'd tested his predictions against the outcomes — how the various candidates had actually performed in officer training. His predictions were worthless. And yet, because it was the army and he had a job to do, he kept on making them; and because he was Danny, he noted that he still felt confident about them. The situation reminded him of the famous Müller-Lyer optical illusion.

Presented with two lines of equal length,

the eye is tricked into seeing one as being longer than the other. Even after you prove to people, with a ruler, that the lines are identical, the illusion persists: They'll insist that one line still looks longer than the other. If perception had the power to overwhelm reality in such a simple case, how much power might it have in a more complicated one?

Danny's commanding officers believed that each branch of the Israel Defense Forces had its own personality. There was a "fighter pilot" type, and an "armored unit" type, and an "infantry soldier" type, and so on. They wanted Danny to determine for which branch any particular recruit was best suited. Danny set out to create a personality test that would effectively sort the entire population of Israel into the correct buckets. He began by listing the handful of traits he thought most obviously correlated with a man's fitness for combat service: masculine pride, punctuality, sociability, sense of duty, capacity for independent thought. "The list of traits was not derived from anything," he later said. "I just thought it up. A professional would take years to do it, using pretests, trying out multiple versions, etcetera, but I didn't know it was difficult to do."

The hard part, Danny thought, was getting an accurate measure of any of these traits from an ordinary job interview. The subtle difficulties that arise when people evaluate other people had been described back in 1915 by an American psychologist named Edward Thorndike. Thorndike asked U.S. Army officers to rate their men according to some physical trait ("physique," for example) and then assess some less tangible quality ("intelligence," "leadership," and so forth). He discovered that the feeling created by making the first ranking bled into the second: If an officer thought a soldier physically impressive, he also found him impressive in other ways. Switch the order of assessment, and the same problem occurred: If a person was first judged to be generally great, he was then judged to be stronger than he actually was. "Obviously a halo of general merit is extended to influence the rating for the special ability, or vice versa," Thorndike concluded; he went on to say that he had "become convinced that even a very capable foreman, employer, teacher, or department head is unable to view an individual as a compound of separate qualities and to assign a magnitude to each of these in independence of the others." Thus was born what is still called "the halo effect."

Danny knew of the halo effect. And he could see that the Israeli army interviewers had been its victims: They had been spending twenty minutes with each new recruit and from the encounter offering a general impression of the recruit's character. General impressions had been proven to be misleading, and so Danny wanted to avoid them. For that matter, he wanted to avoid having to rely on human judgment. Exactly why he mistrusted human judgment he was unsure. In retrospect, he suspected he must have read a recent book by Paul Meehl — the same Meehl who wondered what, if anything, unified the field of psychology. Meehl's book, called *Clinical versus Statistical Prediction,* had shown that psychoanalysts who tried to predict what would become of their neurotic patients fared poorly compared to simple algorithms. Published in 1954 — just a year before Danny overhauled the Israeli army's assessment of the country's youth — it had angered psychoanalysts, who believed that their clinical judgments and predictions had great value. It also raised a more general question: If these putative experts could be misled about the value of their predictions, who would not be misled? "All I know is that I must have read Meehl because of what I did," said Danny.

What he did was teach the army interviewers — young women, mainly — how to put a list of questions to each recruit to minimize the halo effect. He told them to pose very specific questions, designed to determine not how a person thought of himself but how the person had actually behaved. The questions were not just fact-seeking but designed to disguise the facts being sought. And at the end of each section, before moving on to the next, the interviewer was to assign a rating from 1 to 5 that corresponded with choices ranging from "never displays this kind of behavior" to "always displays this kind of behavior." So, for example, when evaluating a recruit's sociability, they'd give a 5 to a person who "forms close social relationships and identifies completely with the whole group" and a 1 to "a person who was "completely isolated." Even Danny could see that there were all kinds of problems with his methods, but he didn't have the time to worry too much about them. For instance, he briefly agonized over how to define a 3 — was it someone who was extremely sociable on occasion, or someone who was moderately sociable all the time? Both, he basically decided. The big thing was that the judge was to keep her private opinions to herself. The question was not "What do I think of him?" but "What has he done?" The judgment of who went where in the Israeli army was to be made by Danny's algorithm. "The interviewers *hated* it," he recalled. "I had a mutiny on my hands. I still remember one of them saying, 'You're turning us into robots.' They had a sense that they could tell [a person's character]. And I was robbing them of that. And they *really* didn't like it."

Danny then had himself driven by an assistant around the country so that he could ask army officers to assign character trait ratings to their own soldiers — which he could then compare to the soldiers' performance. Find the characteristics of the people who are good in a particular branch of the military, his thinking went, and you could use them to identify others who shared those traits and should be assigned to that branch. (His memory of his trip was typically unusual, preserving a curious detail rather than the broad picture. He didn't recall much about his encounters with combat officers, but he remembered vividly what the driver had said after Danny had taken the wheel of the jeep. Danny had never before driven. After he braked in anticipation of a bump in the road, the driver praised him: "He said, 'That is exactly the right gentleness.' ") From the combat officers in the field Danny learned that he'd been sent on a fool's errand. The military stereotypes were false. There were no meaningful differences between the personalities of successful people in the different branches. The personality that succeeded in the infantry was more or less the same as the personality that succeeded beside an artillery piece or inside a tank.

The scores on Danny's personality test did predict something, however. They predicted the likelihood the recruit would succeed *in any job.* They gave the Israeli army a better idea than it had before of who would succeed as an officer, or as a member of some elite service (fighter pilot, paratrooper), and who would not. (They also turned out to predict who would end up in jail.) Maybe more surprisingly, the results were only loosely correlated with intelligence and education — which is to say they contained information that those simple measures did not. The effect of what became known informally as the "Kahneman score" was to make better military use of an entire nation and, in particular, to reduce, in the selection of its military leaders, the importance of raw, measurable intelligence and increase the importance of the qualities on Danny's list.

The process Danny created proved to be so successful that the Israeli military has used it right up to the present day with only minor adjustments. (When women were admitted to combat units, for instance, "masculine pride" became "pride.") "They tried to really change it once," says Reuven Gal, the author of *A Portrait of the Israeli Soldier.* Gal served for five years as chief psychologist of the Israel Defense Forces. "They made it worse, so they changed it back." Upon leaving the army in 1983, Gal went to Washington, DC, on a National Academy of Sciences research associateship. There, one day, he had a call from a top general in the Pentagon. "He says, 'Would you mind coming to talk to us?' " Gal went over to the Pentagon to be interrogated by a roomful of U.S. Army generals. They put their question in many different ways, but, Gal said, "It was always the same question: 'Please explain to me how it is possible you guys use the same rifles we use, drive the same tanks we drive, fly the same airplanes we fly, and you are doing so well winning all of the battles and we are not? I know it's not the weapons. It must be the psychology.

How do you pick the soldiers for combat?' For the next five hours they picked my brain about one thing: our selection process."

Later, when he was a university professor, Danny would tell students, "When someone says something, don't ask yourself if it is true. Ask what it might be true of." That was his intellectual instinct, his natural first step to the mental hoop: to take whatever someone had just said to him and try not to tear it down but to make sense of it. The question the Israeli military had asked him — Which personalities are best suited to which military roles? — had turned out to make no sense. And so Danny had gone and answered a different, more fruitful question: How do we prevent the intuition of interviewers from screwing up their assessment of army recruits? He'd been asked to divine the character of the nation's youth. Instead he'd found out something about people who try to divine other people's character: Remove their gut feelings, and their judgments improved. He'd been handed a narrow problem and discovered a broad truth. "The difference between Danny and the next nine hundred and ninety-nine thousand nine hundred and ninety-nine psychologists is his ability to find the phenomenon and then explain it in a way that applies to other situations," said Dale Griffin, a psychologist at the University of British Columbia. "It looks like luck but he keeps doing it."

A different, more ordinary person would have left the experience brimming with confidence. In a stroke, twenty-one-year-old Danny Kahneman had exerted more influence upon the Israeli army — the institution on which the society depended for its survival — than any psychologist had ever done or ever would do. The obvious next step for him was to go off and get his PhD and become Israel's leading expert in personality assessment and selection processes. Harvard was home to some of the leading figures in the field, but Danny decided, without anyone's help, that he wasn't bright enough to go to Harvard — and didn't bother to apply. Instead he went to Berkeley.

When he returned to Hebrew University as a young assistant professor in 1961, after four years away, he was freshly inspired by personality studies being done by the psychologist Walter Mischel. In the late 1960s Mischel created these wonderfully simple tests on children that wound up revealing a lot about them. In what became known as the "marshmallow experiment," Mischel put three-, four-, and five-year-old kids in a room alone with their favorite treat — a pretzel stick, a marshmallow — and told them that if they could last a few minutes without eating the treat they'd receive a second treat. A small child's ability to wait turned out to be correlated with his IQ and his family circumstances and some other things as well. Tracking the kids through life, Mischel later found that the better a five-year-old resisted the temptation, the higher his future SAT scores and his sense of self-worth, and the lower his body fat and the likelihood he'd suffer from some addiction.

Gripped by a new enthusiasm, Danny designed a bunch of marshmallow test–like experiments. He even coined a phrase for what he was doing: *the psychology of single questions.* He arranged for Israeli kids on camping trips — this was just one example — to be offered a choice between sleeping in a single tent, a two-person tent, or an eight-person tent. Perhaps their answers, Danny thought, would say something about their tendency to affiliate with a group. The idea yielded either no findings or findings he couldn't replicate in a subsequent experiment. And so he gave up. "I wanted to be a scientist," he said. "And I thought, I can't be a scientist unless I can replicate myself. I couldn't replicate myself." Doubting himself once again, he abandoned the study of personality, deciding he had no talent for it.

3
The Insider

Amnon Rapoport was just eighteen years old when he was identified by the Israeli army's new selection system as leadership material. They'd made him a tank commander. "I didn't even know there was a tank corps," he said. One night in October 1956 he drove his tank into Jordan to avenge the murder of several Israeli civilians. On these raids you never knew what decisions you might have to make quickly. Shoot or hold fire? Kill or let live? Live or die? A few months earlier, an Israeli soldier Amnon's age had been captured by the Syrians. He'd decided to kill himself before they could question him. When the Syrians sent his body back, the Israeli army found a note in his toenail: *"I never betrayed."*

On that night in October 1956, Amnon's first decision had been to stop firing: His job was to bombard the second floor of a Jordanian police building until Israeli paratroopers stormed the ground floor. He worried about killing his own men. After he'd stopped shelling he heard, over his tank's radio, reports from the ground. "And all of a sudden, the reality hit me; this was not just an adventure with heroes and villains acting their role. People were dying." The paratroopers were Israel's elite fighting force. Their unit, in hand-to-hand combat, was suffering serious causalities, and yet their reports from the battle to Amnon's ears inside the tank sounded calm, almost casual. "There was no panic," he said, "indeed, no change of intonation and hardly any expression of emotion." These Jews had become Spartans: How had that happened? He wondered how he would fare in hand-to-hand combat. He aspired to be a warrior, too.

Two weeks later he drove his tank into Egypt, in what turned out to be the start of a military invasion. In the fog of battle, he was strafed not just by Egyptian but also Israeli warplanes. His most vivid memory was of an Egyptian MiG-15 diving straight down on his tank while he — with his head above the turret to maintain a 360-degree view of the battlefield — shouted at his driver to zig and zag to avoid being hit. It felt like the MiG was on a special assignment to blow off his head. A few days later, desperate Egyptian soldiers in full retreat approached Amnon's tank with their arms in the air. They begged for water and protection from the Bedouins who hunted them for their rifles and boots. The day before, he was murdering these people; now all he felt toward them was pity. He marveled again — "at how easy it is to shift from an efficient killing machine to compassionate human being, and how quick the switch may be." How did that happen?

After the battles Amnon just wanted to get away from it all. "I was a little bit wild after two years in the tank," he said. "I wanted to go as far away as possible. Flying out of the country was too expensive." Israelis in the 1950s didn't talk about combat stress or its discontents: They just dealt with it. He took a job in a copper mine in the desert just north of the Red Sea — said to be one of the legendary mines of King Solomon. His math skills were better than any of the other workers', most of whom were prison labor, and so he was made the mine's bookkeeper. Among the conveniences that King Solomon's mine was unable to provide was a toilet, or toilet paper. "I went out to take — excuse me — to take a shit. I saw a note in the newspaper that I took to wipe my ass. It said they were opening a psychology department at Hebrew University." He was twenty years old. What he knew of psychology was Freud and Jung — "there were not many textbooks in psychology in Hebrew" — but the subject interested him. He couldn't say why. Nature had called, psychology had answered.

Entrance into Israel's first psychology department, unlike entrance into most Hebrew University departments, was to be competitive. A few weeks after he'd read the ad in the newspaper, Amnon stood in line outside the monastery that served as Hebrew University, waiting to take a series of bizarre tests — including one designed by Danny Kahneman, who had written a page of prose in a language he had invented so that applicants might attempt to decipher its grammatical structure. The line of applicants ran down the block. There were only twenty or so spots in the new department, but hundreds of people wanted into it: An amazing number of young Israelis, in 1957, wanted to know what made people tick. The talent was also incredible: Of the twenty people admitted, nineteen went on to earn their doctorates, and the one who didn't was a woman who, scoring one of the top marks on the admissions test, then had her career derailed by children. Israel without a psychology department was like Alabama without a football team.

In line beside Amnon stood a small, pale, baby-faced soldier. He looked about fifteen but he wore, almost absurdly, the high, rubber-soled boots and crisp uniform and red beret of the Israeli paratrooper. The new Spartan. Then he started to speak. His name was Amos Tversky. Amnon wouldn't remember exactly what he had said but he'd remember, vividly, how he'd felt about it. "I was not as smart as he was. I understood it immediately."

To his fellow Israelis, Amos Tversky somehow was, at once, the most extraordinary person they had ever met and the quintessential Israeli. His parents were among the pioneers who had fled Russian anti-Semitism in the early 1920s to build a Zionist nation. His mother, Genia Tversky, was a social force and political operator who became a member of the first Israeli Parliament, and the next four after that. She sacrificed her private life for public service and didn't agonize greatly about the choice. She was often gone — she spent two years of Amos's early childhood in Europe, helping the U.S. Army liberate the concentration camps and resettle the survivors. Upon her return she spent more time at the Knesset in Jerusalem than at home.

Amos had a sister, but she was thirteen years older, and he was raised, in effect, as an only child. The person who did most of that raising was his father, a veterinarian who spent much of his time treating livestock. (Israelis couldn't afford pets.) Yosef Tversky, the son of a rabbi, despised religion and loved Russian literature, and found a great deal of amusement in what came out of the mouths of his fellow human beings. His father had turned away from an early career in medicine, Amos explained to friends, because "he thought animals had more real pain than people and complained a lot less." Yosef Tversky was a serious man. At the same time, when he talked about his life and work, he brought his son to his knees with laughter about his experiences, and about the mysteries of existence. "This work is dedicated to my father, who taught me to wonder," Amos would one day write at the opening of his PhD dissertation.

Amos was fond of saying that interesting things happened to people who could weave them into interesting stories. He, too, could tell a story, with startlingly original effect. He spoke with a slight lisp that reminded some of the way that Catalans spoke Spanish. He was so pale that his skin was almost translucent. Whether he was speaking or listening, his pale blue eyes darted back and forth, as if searching for an approaching thought.

Even as he spoke, he gave the impression of constant motion. He wasn't conventionally athletic — he was always small — but he was loose-jointed and fast: twitchy and incredibly agile. He had an almost feral ability to run at great speed up and down mountains. One of his favorite tricks — he'd sometimes do this as he told a story — was to place himself on a high surface, whether a rock or a table or an army tank, and fall face-first toward the ground. His body perfectly horizontal to the earth, he'd fall until people shrieked and then he'd pull himself up at the last moment and somehow land on his feet. He loved the sensation of falling, and the view of the world from above.

Amos was also physically brave, or at least intent on seeming so. Not long after his parents moved him from Jerusalem, in 1950, to the coastal city of Haifa, he found himself at a swimming pool with other kids. The pool had a ten-meter diving platform. The kids challenged him to jump off it.

Amos was twelve years old but didn't yet know how to swim. In Jerusalem, during the war of independence, they hadn't had water to drink, much less to fill swimming pools with. Amos found a big kid and said, *I'm going to do this, but I need you to be in the pool when I land, to pull me up from the bottom.* Amos jumped, and the big kid rescued him from drowning and pulled him out of the pool.

Entering high school, Amos, like all Israeli kids, needed to decide if he would specialize in math and science or in the humanities. The new society exerted great pressure on boys to study math and science. That's where the status was, and the future careers. Amos had a gift for math and science, perhaps more than any other boy. And yet alone among the bright boys in his class — and to the bemusement of all — he pursued the humanities. Another risky leap into the unknown: He could teach himself math, Amos said, and he couldn't ignore the thrill of studying with the humanities teacher, a man named Baruch Kurzweil. "In contrast to most of the teachers, who spread boredom and superficiality, I'm full of enjoyment and amazement in his classes in Hebrew literature and philosophy," Amos wrote to his older sister Ruth, who had moved to Los Angeles. Amos wrote poetry for Kurzweil and told people he planned to become a poet or a literary critic.*

He formed an intense, private, possibly romantic relationship with a new student named Dahlia Ravikovitch. She'd turned up one day, morosely, in their high school class. After her father's death she'd lived on a kibbutz, which she loathed, then bounced unhappily through a series of foster homes. She was the picture of social alienation, or at any rate the 1950s Israeli version of it, and yet Amos, the most popular kid in the school, took up with her. The other kids didn't know what to make of it. Amos still looked like a boy; Dahlia seemed, in every

* When B. F. Skinner discovered as a young man that he would never write the great American novel, he felt a despair that he claimed nearly drove him into psychotherapy. The legendary psychologist George Miller claimed that he gave up his literary ambition for psychology because he had nothing to write about. Who knows what mixed feelings William James experienced when he read his brother Henry's first novel? "It would be interesting to ask how many psychologists come up short next to great writers who happen to be near them," one prominent American psychologist has said. "It may be the fundamental driver."

way, already a grown woman. He loved the outdoors and games; she . . . well, when all the other girls went out to gym class, she sat at the window and smoked. Amos loved being with big groups of people; Dahlia was a loner. It was only later, when Dahlia's poetry claimed Israel's highest literary prizes and she became a global sensation, that people said, "Oh, that made sense. Two geniuses." Just as it made sense, after Baruch Kurzweil became Israel's most prominent literary critic, that Amos had wanted to study with him. But it did, and it didn't. Amos was the most insistently upbeat person anyone knew. Dahlia, like Kurzweil, attempted suicide. (Kurzweil succeeded.)

Like a lot of the Jewish kids in Haifa in the early 1950s, Amos joined a leftist youth movement called the Nahal. He was soon elected a leader. The Nahal — the word was an acronym for the Hebrew phrase meaning "Fighting Pioneer Youth" — was a vehicle to move young Zionists from school onto kibbutzim. The idea was that they would serve as soldiers and guard the farm for a couple of years and then become farmers.

During Amos's final year in high school, the swashbuckling Israeli general Moshe Dayan came to Haifa to speak to the students. A boy who happened to be in the audience recalls, "He says all those who go to the Nahal, raise your hands? A huge number did. Dayan says, 'You are traitors. We don't want you growing tomatoes and cucumbers. We want you *fighting.*' " The next year every youth group in Israel was asked to pick twelve kids out of every hundred to serve their country not as farmers but paratroopers. Amos looked more like a boy scout than an elite soldier, but he volunteered immediately. Too light to qualify, he drank water until he made weight.

At paratrooper school Amos and the other young men were turned into symbols of the new country: warriors and killing machines. Cowardice wasn't an option. Once they'd proven that they could jump to the ground from a height of eighteen feet without breaking anything, they were taken up in old World War II planes built of wood. The propeller was at the same level as the door but just in front of it, so there was this strong gust of wind to throw you backward the moment you stepped out. The light on the door was red. They checked each other's equipment until the light turned green, and, one by one, they moved forward: Anyone who hesitated was pushed out.

The first few jumps, a lot of the young

men hesitated; they needed a little push. One kid in Amos's group refused to jump and was ostracized for the rest of his life. ("It took *real* bravery not to jump," a former paratrooper later said.) Amos never hesitated. "He was always on the extreme end of enthusiastic when it came to jumping out of airplanes," recalls fellow paratrooper Uri Shamir. He jumped fifty times, maybe more. He jumped behind enemy lines. He jumped into battle in 1956, in the Sinai campaign. Once, he jumped by accident into a hornet's nest and was stung so badly he passed out. After university, in 1961, he flew for the first time in his life without a parachute, to graduate school in the United States. As his plane descended, he looked at the earth below with genuine curiosity, turned to the person sitting beside him, and said, "I've never landed."

Soon after he joined the paratroopers, Amos became a platoon commander. "It is amazing how quickly one is able to adapt to a new way of life," he wrote to his sister in Los Angeles. "The boys my age were no different than I was other than the two stripes on my arm. Now they salute me and follow my every command: to run and to crawl. And now this relationship is accepted, even by me, and seems natural to me." The letters Amos wrote home were censored and offer only a glimpse of his combat experience. He was sent on reprisal missions, which invited atrocities on both sides. He lost men, and saved them. "During one of our 'payback missions,' I saved one of my soldiers and received honorable mention," he wrote to his sister. "I did not think I had done anything heroic, I just wanted my soldiers to return home safely."

There were other ordeals, of which he did not write, and seldom spoke. A sadistic senior Israeli officer wanted to test how far men could travel without their usual provisions and deprived them of water for great stretches. The experiment ended when one of Amos's men died of dehydration; Amos testified against his commanding officer at the latter's court-martial. One night Amos's men threw a blanket over another sadistic officer's head and beat him savagely. Amos didn't join in the beating, but in the subsequent investigation, he helped the men who had done it avoid prosecution. "When they ask you questions, just bore them with lots of irrelevant details and they will be thrown off the scent," he told them, and it had worked.

By late 1956, Amos was not merely a

platoon commander but a recipient of one of the Israeli army's highest awards for bravery. During a training exercise in front of the General Staff of the Israel Defense Forces, one of his soldiers was assigned to clear a barbed wire fence with a bangalore torpedo. From the moment he pulled the string to activate the fuse, the solider had twenty seconds to run for cover. The soldier pushed the torpedo under the fence, yanked the string, fainted, and collapsed on top of the explosive. Amos's commanding officer shouted for everyone to stay put — and leave the unconscious soldier to die. Amos ignored him and sprinted from behind the wall that served as cover for his unit, grabbed the soldier, picked him up, hauled him ten yards, tossed him on the ground, and threw himself on top of him. The shrapnel from the explosion remained in Amos for the rest of his life. The Israeli army did not bestow honors for bravery lightly. As he handed Amos his award, Moshe Dayan, who had watched the entire episode, said, "You did a very stupid and brave thing and you won't get away with it again."

Occasionally, people who watched Amos in action sensed that he was more afraid of being thought unmanly than he was actually brave. "He was always very gung ho," recalled Uri Shamir. "I thought it was maybe compensation for being thin and weak and pale." At some point it didn't matter: He compelled himself to be brave until bravery became a habit. And as his time in the army came to an end he clearly sensed a change in himself. "I cannot rid myself of the feeling that you would almost not know me today," Amos wrote to his sister. "Letters cannot convey the drastic changes of a boy in an army uniform that you will meet. He will be very different from the young boy in khaki shorts that you left at the airport five years ago."

Apart from that short note, Amos seldom mentioned his army experiences, in print or conversation, unless it was to tell a funny or curious story — how, for instance, during the Sinai campaign, his battalion captured a train of Egyptian fighting camels. Amos had never ridden a camel, but when the military operation ended, he won the competition to ride the lead camel home. He got seasick after fifteen minutes and spent the next six days walking the caravan across the Sinai.

Or how his soldiers, even in combat, refused to wear their helmets, claiming that the weather was too hot for them and "if a bullet is going to kill me, it has my name on it anyway." (To which Amos said, "What about all those bullets addressed 'To Whom It May Concern'?") More typically Amos's stories began with some off-hand observation of the world around him. "Almost always when he encountered you he would start the conversation with 'Did I tell you this story?' " recalls Samuel Sattath, an Israeli mathematician. "But the stories were not about him. He would say, for example, 'You know, in an Israeli university meeting, everyone jumps in to speak, because they think someone else might be about to say what they want to say. And in an American university faculty meeting, everyone is quiet, because they think someone else will think to say what they want to say . . .' " And he'd be off on a disquisition on the differences between Americans and Israelis — how Americans believed tomorrow will be better than today, while Israelis were sure tomorrow would be worse; how American kids always came to class prepared, while Israeli kids never did the reading, but it was Israeli kids who always had the bold idea, and so on.

To those who knew Amos best, Amos's stories were just an excuse to enjoy Amos. "People who knew Amos could talk of nothing else," as one Israeli woman, a friend of long standing, put it. "There was nothing we liked to do more than to get together and talk about him, over and over and over." There were — for starters — the stories about the funny things Amos had said, usually directed at people whom he found full of themselves. He had listened to an American economist talk about how so-and-so was stupid and so-and-so was a fool, then said, "All your economic models are premised on people being smart and rational, and yet all the people you know are idiots." He'd heard Murray Gell-Mann, a Nobel laureate in physics, hold forth on seemingly every subject under the sun. After Gell-Man was done, Amos said, "You know, Murray, there is no one in the world who is as smart as you think you are." Once, after Amos gave a talk, an English statistician had approached him. "I don't usually like Jews but I like you," the statistician said. Amos replied, "I usually like Englishmen but I don't like you."

The effect on others of whatever Amos said only led to even more stories about Amos. There was — to take just one example — the time that Tel Aviv University threw a party for a physicist who had just won the Wolf Prize. It was the discipline's second-highest honor, and its winners more often than not went on to win the Nobel.

Most of the leading physicists in the country came to the party, but somehow the prizewinner ended up in the corner with Amos — who had recently taken an interest in black holes. The next day the prizewinner called his hosts to ask, "Who was that physicist I was talking to? He never told me his name." After some confusing back-and-forth, his hosts figured out that the man meant Amos, and they told him that Amos wasn't a physicist but a psychologist. "It's not possible," the physicist said, "he was the smartest of all the physicists."

The Princeton philosopher Avishai Margalit said, "No matter what the topic was, the first thing Amos thought was in the top 10 percent. This was such a striking ability. The clarity and depth of his first reaction to any problem — any intellectual problem — was something mind-boggling. It was as if he was right away in the middle of any discussion." Irv Biederman, a psychologist at the University of Southern California, said, "Physically he was unremarkable. In a room full of thirty people he'd be the last one you'd notice. And then he'd start to talk. Everyone who ever met him thought he was the smartest person they had ever met." The University of Michigan psychologist Dick Nisbett, after he'd met Amos, designed a one-line intelligence test: The sooner you figure out that Amos is smarter than you are, the smarter you are. "He would walk into a room," recalled his close friend and collaborator Varda Liberman, a mathematician. "He didn't look special. And the way he dressed said nothing. He'd sit there quietly. And then he would open his mouth and speak. And in no time he became the light that all the butterflies fly to; and in no time everyone would look up to him wanting to hear what he would say."

Even so, most of the stories people told about Amos had less to do with what came out of his mouth than with the unusual way he moved through the world. He kept the hours of a vampire. He went to bed when the sun came up and woke up at happy hour. He ate pickles for breakfast and eggs for dinner. He minimized quotidian tasks he thought a waste of time — he could be found in the middle of the day, having just woken up, driving himself to work while shaving and brushing his teeth in the rear-view mirror. "He never knew what time of the day it was," said his daughter, Dona. "It didn't matter. He's living in his own sphere and you just happened to encounter him there." He didn't pretend to be interested in whatever others expected him to be interested in — God help anyone who tried to drag him to a museum or a board meeting. "For those who like that sort of thing, that is the sort of thing they like," Amos liked to say, plucking a line from the Muriel Spark novel *The Prime of Miss Jean Brodie.* "He just skipped family vacations," says his daughter. "He'd come if he liked the place. Otherwise he didn't." The children didn't take it personally: They loved their father and knew that he loved them. "He loved people," said his son Oren. "He just didn't like social norms."

A lot of things that most human beings would never think to do, to Amos simply made sense. For instance, when he wanted to go for a run he . . . went for a run. No stretching, no jogging outfit or, for that matter, jogging: He'd simply strip off his slacks and sprint out his front door in his underpants and run as fast as he could until he couldn't run anymore. "Amos thought people paid an enormous price to avoid mild embarrassment," said his friend Avishai Margalit, "and he himself decided very early on it was not worth it."

What all those who came to know Amos eventually realized was that the man had a preternatural gift for doing only precisely what he wanted to do. Varda Liberman recalled visiting him one day and seeing a table with a week's worth of mail on it. There were tidy little stacks, one for each day, each filled with requests and entreaties and demands upon Amos's time: job offers, offers of honorary degrees, requests for interviews and lectures, requests for help with some abstruse problem, bills. When the new mail came in Amos opened anything that interested him and left the rest in its daily pile. Each day the new mail arrived and shoved the old mail down the table. When a pile reached the end of the table Amos pushed it, unopened, off the edge into a waiting garbage can. "The nice thing about things that are urgent," he liked to say, "is that if you wait long enough they aren't urgent anymore." "I would say to Amos I have to do this or I have to do that," recalled his old friend Yeshu Kolodny. "And he would say, 'No. You don't.' And I thought: lucky man!"

There was this beautiful simplicity to Amos: His likes and dislikes could be inferred directly and accurately and at all times from his actions. Amos's three children have vivid memories of watching their parents drive off to see some movie picked by their mother, only to have their father turn up back at their couch twenty minutes later. Amos would have decided, in the first five minutes, whether the movie was worth seeing — and if it wasn't he'd just come home and watch *Hill Street Blues* (his favorite TV drama) or *Saturday Night Live* (he never missed it) or an NBA game (he was obsessed with basketball). He'd then go back and fetch his wife after her movie ended. "They've already taken my money," he'd explain. "Should I give them my time, too?" If by some freak accident he found himself at a gathering of his fellow human beings that held no appeal for him, he'd become invisible. "He'd walk into a room and decide he didn't want anything to do with it and he would fade into the background and just vanish," says Dona. "It was like a superpower. And it was absolutely an abnegation of social responsibility. He didn't accept social responsibility — and so graciously, so elegantly, didn't accept it."

Occasionally Amos offended someone — of course he did. His darting pale blue eyes were enough to unsettle people who didn't know him. Their constant motion gave them the impression he wasn't listening to them, when the problem, often, was that he had listened too well. "For him the main thing is the people who don't know the difference between knowing and not knowing," says

Avishai Margalit. "If he thought you were a bore and there was nothing there, he could cut you like nothing." Those who knew him best learned how to rationalize whatever he had said or done.

It never occurred to him that anyone with whom he wanted to spend time wouldn't want to spend time with him. "He expected first of all to charm you," said Samuel Sattath. "Which was odd for such a smart person." "He sort of invited people to love him," said Yeshu Kolodny. "When you were on the good side of Amos he was very easy to love. Extremely easy. There was a competition around him. People competed for Amos." It was a very common thing for Amos's friends to ask themselves: *I know why I like him, but why does he like me?*

Amnon Rapoport did not lack for admirers. He'd been famously brave in battle. Israeli women, taking in for the first time his blond hair and tanned skin and chiseled features, often decided he was the best-looking man they'd ever laid eyes on. Onc day he'd earn his PhD in mathematical psychology and become a highly regarded professor, with his pick of the world's universities. And yet he, too, when he sensed that Amos liked him, wondered why. "I know that what attracted me to Amos was how clever he was," said Amnon. "I don't know what attracted him to me. I was supposed to be very handsome, maybe that." Whatever its source, the attraction was strong. From the moment they met, Amnon and Amos were inseparable. They sat side by side in the same classes; they lived in the same apartments; they spent summers hiking the country together. They were famously a pair. "I think some people thought we were homosexual or something," said Amnon.

Amnon also had the best seat in the house when Amos decided what he was going to do with his life. Hebrew University in the late 1950s required students to pick two fields of concentration. Amos had chosen philosophy and psychology. But Amos approached intellectual life strategically, as if it were an oil field to be drilled, and after two years of sitting through philosophy classes he announced that philosophy was a dry well. "I remember his words," recalled Amnon. "He said, 'There is nothing we can do in philosophy. Plato solved too many of the problems. We can't have any impact in this area. There are too many smart guys and too few problems left, and the problems have no solutions.' " The mind-body problem was a good example. How are our various mental events — what you believe, what you think — related to our physical states? What is the relationship between our bodies and our minds? The question was at least as old as Descartes, but there was still no answer in sight — at least not in philosophy. The trouble with philosophy, Amos thought, was that it didn't play by the rules of science. The philosopher tested his theories of human nature on a sample size of one — himself. Psychology at least pretended to be a science. It kept at least one hand at all times on hard data. A psychologist might test whatever theory he devised on a representative sample of humanity. His theories might be tested by others, and his findings reproduced, or falsified. If a psychologist stumbled upon a truth he might make it stick.

To Amos's closest Israeli friends, there was never anything mysterious about his interest in psychology. Questions of why people behaved as they behaved, and thought as they thought, were thick in the air they breathed. "You never discussed art," recalls Avishai Margalit. "You discussed people. It was a constant thing, a constant puzzle: What makes others tick? It comes from the shtetl. Jews were petty merchants. They had to assess others, all the time. Who is dangerous? Who is not dangerous? Who will repay the debt, who won't repay the debt? People were basically dependent on their psychological judgment." Still, to many, the presence of a mind as clear as Amos's in a field as murky as psychology remained a mystery. How had this relentlessly optimistic person, with his clear and logical mind and zero tolerance for bullshit, wound up in a field cluttered with unhappy souls and mysticism?

Amos, when he talked about it, which he usually didn't, made it seem as if it had started as a whim. When he was in his mid-forties and many of the brightest young minds in the field wanted to study with him, he sat down with a professor of psychiatry at Harvard named Miles Shore. Shore asked him how he had become a psychologist. "It's hard to know how people select a course in life," Amos said. "The big choices we make are practically random. The small choices probably tell us more about who we are. Which field we go into may depend on which high school teacher we happen to meet. Who we marry may depend on who happens to be around at the right time of life. On the other hand, the small decisions are very systematic. That I became a psychologist is probably not very revealing.

What *kind* of psychologist I am may reflect deep traits."

What kind of psychologist would he be? In most of psychology Amos found little to interest him. After taking classes in child psychology and clinical psychology and social psychology, he concluded that the vast majority of his chosen field was safely ignorable. To his assigned work he paid shockingly little attention. His classmate Amia Lieblich witnessed Amos's insouciance after he'd been assigned by a professor to administer an intelligence test to a five-year-old child. "The night before the work was due, Amos turned to Amnon and said, 'Amnon, lie down on the couch. I am going to ask you some questions. Pretend you are five years old.' And he got away with it!" Amos was the only student who never took notes in class. When the time came to study for some test, Amos would simply ask to see Amnon's notes. "He would read my notes once and know the material better than I did," said Amnon. "It was the same way he could meet a physicist in the street, talk to him for thirty minutes, without knowing anything about physics, and then tell the physicist something about physics the physicist didn't know. I first thought he was a superb superficial person — that it was a party trick. And that was a mistake. Because it wasn't a trick."

It didn't help that so many of the professors seemed to be flying by the seat of their pants. The guy who had come from Scotland to teach the history of psychology was sent back when it was discovered he had fabricated his PhD. A guy they brought in to teach a class on personality testing — a Polish Jew who had survived the Holocaust by hiding in the woods — fled the classroom in tears under questioning from Amos and Amnon. "We basically had to teach psychology to ourselves," recalled Amnon. Amos compared clinical psychology — everywhere on the rise, and the field of greatest interest to their fellow students, most of whom hoped to become therapists — to medicine. If you went to a doctor in the seventeenth century, you were worse off for having gone. By the end of the nineteenth century, going to the doctor was a break-even proposition: You were as likely to come away from the visit better off as you were to be worse off. Amos argued that clinical psychology was like medicine in the seventeenth century, and he had lots of evidence to support his case.

One day during their second year at Hebrew University, in 1959, Amnon came across a paper called "The Theory of Decision Making," by a psychology professor at Johns Hopkins named Ward Edwards. "Many social scientists other than psychologists try to account for the behavior of individuals," it opened. "Economists and a few psychologists have produced a large body of theory and a few experiments that deal with individual decision making. The kind of decision making with which this body of theory deals is as follows: given two states, *A* and *B,* into either one of which an individual may put himself, the individual chooses *A* in preference to *B* (or vice versa). For instance, a child standing in front of a candy counter may be considering two states. In state *A* the child has $0.25 and no candy. In state *B* the child has $0.15 and a ten-cent candy bar. The economic theory of decision making is a theory about how to predict such decisions." Edwards went on to lay out a problem: Economic theory, the design of markets, public policy making, and a lot more depended on theories about how people made decisions. But psychologists — the people most likely to test these theories and determine how people actually made decisions — hadn't paid much attention to the subject.

Edwards wasn't setting himself, or his

field, in opposition to economics. He was merely proposing that psychologists be invited, or perhaps invite themselves, to test both the assumptions and the predictions made by economists. Economists assumed that people were "rational." What did they mean by that? At the very least, they meant that people could figure out what they wanted. Given some array of choices, they could order them logically, according to their tastes. For example, if they were handed a menu that listed three hot drinks, and they said that at some given moment they preferred coffee to tea, and tea to hot chocolate, they should logically prefer coffee to hot chocolate. If they preferred *A* to *B* and *B* to *C*, they should prefer *A* to *C*. In the academic jargon, they were "transitive." If people couldn't order their preferences logically, how would any market ever function properly? If people preferred coffee to tea and tea to hot chocolate — but then turned around and chose hot chocolate over coffee — they'd never finish choosing. They'd be willing, in principle, to pay to switch from hot chocolate to tea and also to switch from tea to coffee — and then pay again to switch from coffee to hot chocolate. They'd never settle on a drink but instead would be stuck in this mad infinite loop in which they kept paying to upgrade from the drink they had to a drink they liked better.

Here was one of the predictions that economists made that Edwards thought psychologists might test: Are actual human beings transitive? If at any given moment they preferred coffee to tea and tea to hot chocolate, did they prefer coffee to hot chocolate? A few people had recently looked into the matter, Edwards noted, among them a mathematician named Kenneth May. Writing in a leading economics journal, *Econometrica,* May described how he had tested just how logical his own students were when asked to choose a spouse. He'd presented students with three potential mates, ranked by three qualities: how good-looking they were, how smart they were, and how much money they had. None of the three potential mates was extreme in any one way: No one was so poor, dumb, or hard on the eye as to be repugnant. Each had relative strengths and weaknesses: Each ranked highest in one category, second highest in another, and last in the third. May's students, in making their choices, never faced all three potential marriage partners at the same time. Instead they were shown pairs, and asked to choose between them. For example, they might be asked to choose between the potential mate who was the brightest, second-best-looking, but poorest, and the potential mate who was the richest, the second-brightest, but the least good-looking.

Once the dust had settled in this flurry of decision making, more than a quarter of the students had revealed themselves as irrational, at least from the point of view of economic theory. They'd decided that they would rather marry Jim than Bill, and Bill than Harry — but then also said that they would rather marry Harry than Jim. If people could buy and sell spouses like hot drinks, some large number of them would never settle on one spouse but would instead keep paying to upgrade. Why? May didn't offer a full explanation, but he suggested the beginning of one: Because Jim and Bill and Harry each had relative strengths and weaknesses, they were hard to compare. "It is just these non-comparable cases that are of interest," wrote May. "Comparison of alternatives in which one is superior to the other in every respect makes for a simple but rather trivial theory."

Amnon showed Ward Edwards's paper on decision making to Amos, and Amos grew very excited. "Amos will smell gold before anyone else will smell it," said Amnon. "And he smelled gold."

In the fall of 1961, a few weeks after Amnon flew to the University of North Carolina, Amos left Jerusalem for the University of Michigan — where Ward Edwards had moved after being fired by Johns Hopkins, supposedly for not bothering to show up for the classes he was meant to be teaching. Neither Amnon nor Amos knew much about American universities. Amnon, who had just been assigned to North Carolina by a Fulbright scholarship committee, had to pull out his *Atlas of the World* to find it. Amos was able to read English, but he spoke so little that, when he told people where he planned to go, they assumed he was joking. "How will he even survive?" his friend Amia Lieblich asked herself. Neither Amnon nor Amos saw that they had any real choice. "There was nobody to teach us at Hebrew University," Amnon said. "We had to leave." Both Amnon and Amos assumed that the move was temporary: They would learn whatever there was to learn about this new field of decision making in the United States and then return to Israel and work together.

The earliest sightings of Amos Tversky in the United States are anomalies in the History of Amos. In their first week of classes, fellow students saw a silent, seemingly dutiful foreigner taking notes. They looked upon him with pity. "My first memory is of him being really, really quiet," recalls fellow graduate student Paul Slovic. "Which is funny, because later on he *really* wasn't quiet." Seeing Amos writing from right to left, one student suggested that he might suffer from some mental disorder. (He was writing in Hebrew.) Stripped of the power of speech, Amos was jolted out of character. Long after the fact, Paul Slovic guessed that in his first few months away from home Amos merely had been biding his time. Until he knew exactly what he was saying, he wouldn't say it.

By the middle of his first year Amos knew what he was saying — and from that moment the Amos stories came thick and fast. There was the time that Amos walked into an Ann Arbor diner and ordered a hamburger with relish. The waiter said they didn't have relish. Okay, Amos said, I'll have tomato. We don't have tomato, either, said the waiter. "Can you tell me what else you don't have?" asked Amos. There was the time Amos had arrived late for what everyone expected would be a grueling test, given by a dreaded professor of statistics, John

Milholland. Amos slid into a desk just as the test was being passed out. The room was dead silent, the students anxious and tense. As Milholland reached his desk, Amos turned to the person seated next to him and said, "Forever and forever, farewell, John Milholland / If we do meet again, why, we shall smile / If not, why then, this parting was well made": lines spoken by Brutus to Cassius in act 5, scene 1, of *Julius Caesar.* He aced the test.

Michigan required that all PhD students in psychology pass a proficiency test in two foreign languages. Weirdly, the university didn't count Hebrew as a foreign language but accepted mathematics. Though entirely self-taught in mathematics, Amos chose math as one of his languages and passed the test. For his second language he picked French. The test was to translate three pages from a book in the language: The student chose the book, and the tester chose the pages to translate. Amos went to the library and dug out a French math textbook with nothing but equations in it. "It might have had the word *donc* in it," said Amos's roommate Mel Guyer. The University of Michigan declared Amos Tversky proficient in French.

Amos wanted to explore how people made

decisions. To do this he required subjects who were both captive and poor enough that they would respond to the tiny financial incentives he could offer. He found them in the maximum security wing of the Jackson State Prison, near Ann Arbor. Amos offered the inmates — though only those with IQs over 100 — different gambles, involving candy and cigarettes. Both functioned in the jail as currency, and everyone knew what they were worth — a pack of cigarettes and a sack of candy at the prison store each cost 30 cents, or about a week's salary. The inmates could either take the gamble or sell the right to take the gamble to Amos — that is, receive a sure payout.

As it turned out, the Jackson Prison inmates choosing between gambles had a lot in common with Kenneth May's students when they chose between spouses: After they had said they preferred *A* to *B* and *B* to *C*, they could be induced to prefer *C* to *A*. Even when you asked them up front whether they would ever chose *C* over *A* and they insisted they would never do such a thing, they did it. Some thought Amos must be playing a trick on the inmates, but he wasn't. "He didn't trick the prisoners into violating transitivity," says Michigan professor Rich Gonzalez. "He used a process much like the old saying about the frog in the pot of boiling water. As the temperature increases slowly, the frog can't detect it. Obviously the frog can detect 90 degrees versus 200 degrees, but not increments of a single degree. In some of our biological systems we are equipped to detect big differences; in others, small ones — say, a tickle versus a poke. If people can't detect small differences, Amos figured, they might violate transitivity."

Clearly people had trouble detecting small differences. Prison inmates and Harvard students, on whom Amos also ran tests. He wrote a paper about his experiments in which he showed how one might even predict when people would be intransitive. And yet . . . he didn't read much into this. Rather than draw some grand conclusions about the inadequacy of existing assumptions about human rationality, he pulled himself up short. "Is this behavior irrational?" he wrote. "We tend to doubt it. . . . When faced with complex multidimensional alternatives, such as job offers, gambles or [political] candidates, it is extremely difficult to utilize properly all the available information." It wasn't that people actually preferred *A* to *B* and *B* to *C* and then turned around and preferred *C* to *A*. It was that it was sometimes very hard to understand the differences. Amos didn't think that the real world was as likely to fool people into contradicting themselves as were the experiments he had designed.

The man whose work had pulled Amos to Michigan, Ward Edwards, turned out to be more appealing to Amos on the page than in the flesh. After Johns Hopkins fired him, Edwards found a place in Michigan, but his position was insecure, and so was he. When students arrived to work with him, he gave each of them a pompous little lecture — they called it the "key" lecture. Edwards would hold up the key to the door of the small house that served as his lab and tell the student what an honor it was for him to be entrusted with the key and, by extension, an association with Edwards. "You got this key along with the speech," says Paul Slovic. "The meaning of the key, the symbol of the key — it was all a little weird. Usually someone just gives you a key and tells you to make sure you lock the door when you leave."

Edwards hosted a party at his house for some visiting scholar — and charged his guests for the beer. He sent Amos out to do research for him and then withheld his expenses until Amos put up a fight. He insisted that any work Amos did in his lab was at least in part the property of Ward Edwards, and thus any paper that Amos wrote should also have Ward Edwards's name on it. Amos liked to say that stinginess was contagious and so was generosity, and since behaving generously made you happier than behaving stingily, you should avoid stingy people and spend your time only with generous ones. He paid attention to what Edwards was up to without paying a lot of attention to Edwards himself.

The University of Michigan was then, as it is now, home to the world's largest department of psychology. There were others in it thinking about decision making, and Amos found himself drawn to one of them, Clyde Coombs. Coombs drew a distinction between the sorts of decisions in which more was better, and more subtle decisions. For instance, other things being equal, just about everyone would decide to take more money rather than less, and to accept less pain rather than more. What interested Coombs were the fuzzier decisions. How does a person decide where to live, or whom to marry, or, for that matter, which jam to buy? The giant food company General Mills had hired Coombs in hopes that he might create for them tools to measure their customers' feelings about their products. But how do you measure the strength of a person's feelings for Cheerios? What kind of scale do you use? A person might be twice as tall as another person, but might he like something twice as much? One place might be ten degrees hotter than another place; could one person's feelings for a breakfast cereal be ten degrees hotter than another's? To predict what people would decide, you had to be able to measure their preferences: but how?

Coombs thought about the problem first by framing decisions as a series of comparisons between two things. In the mathematical model he built, the choice between, say, two potential spouses became a multistage process. A person had in mind some ideal spouse — or a set of traits that he wanted in a spouse. He compared each of the real-world choices of spouse to the ideal, and chose the spouse who most closely resembled the ideal. Coombs obviously didn't think that, when people chose something, they actually did any such thing. He didn't know what they did. He was just trying to build a tool that would help to predict what human beings would choose when faced with an array of things to choose from. To explain what he was up to — and probably to make it seem less preposterous — Coombs used the example of a cup of tea. How did a person decide how much sugar to put in his tea? Well, he had some notion of the ideal sweetness of tea; he sugared his tea until it most closely resembled that ideal. A lot of life decisions, Coombs thought, were like that, only more complicated.

Take the decision of whom to marry. Presumably people held in their minds at least some vague notion of an ideal spouse — a set of traits they thought important, though perhaps all not equally so — and then chose a person from the available pool who most closely resembled that ideal. To understand the decision, you obviously needed to figure out how much weight people placed on various traits. To a man in search of a wife, how important is intelligence versus looks? Or looks versus personal finances? You also needed to figure out how people assessed those traits in the first place — how a woman seeking a husband, say, compared her notional ideal of a husband to the man she has just met. How on earth does a woman decide how similar the sense of humor of the guy sitting across the speed dating table from her is to her ideal sense of humor? Our decisions, Clyde

Coombs thought, might be treated as a collection of judgments about the similarity between two things: the ideal in our head, and the object on offer.

Amos was as fascinated as Coombs by questions of how to measure what couldn't be observed (so interested that he taught himself the math he needed to do it). But he also saw that the attempt to measure these preferences raised another question. If you were going to take as your (possibly unrealistic) working assumption the proposition that people made choices by comparing some ideal in their head and the real-world versions, you had to know how people made such judgments. "Similarity judgments," psychologists called them, in a rare example of comprehensible trade jargon. What goes on in the mind when it evaluates how much one thing is like, or not like, another? The process is so fundamental to our existence that we scarcely stop to think about it. "It's the process that grinds away constantly and generates much of our understanding and response to the world," says Berkeley psychologist Dacher Keltner. "First of all it's, how do you categorize things? And that's everything. Do I sleep with him or not? Do I eat this or not? Do I give to this person or not? Is that a boy or a girl? Is that a predator or prey? If you solve how the process works, you solve how we know things. It's how knowledge about the world is organized. It's like the thread that is woven thorough everything in the mind."

The reigning theories in psychology of how people made judgments about similarity all had one thing in common: They were based on physical distance. When you compare two things, you are asking how *closely* they resemble each other. Two objects, two people, two ideas, two emotions: In psychological theory they existed in the mind as they would on a map, or on a grid, or in some other physical space, as points with some fixed relationship to each other. Amos wondered about that. He'd read papers by Berkeley psychologist Eleanor Rosch, who in the early 1960s was exploring how people classified objects. What makes a table a table? What makes a color its own distinctive color? In her work, Rosch had asked her subjects to compare colors and judge how similar they were to each other.

People said some strange things. For instance, they said that magenta was similar to red, but that red wasn't similar to magenta. Amos spotted the contradiction and set out to generalize it. He asked people if they thought North Korea was like Red China. They said yes. He asked them if Red China was like North Korea — and they said no. People thought Tel Aviv was like New York but that New York was not like Tel Aviv. People thought that the number 103 was sort of like the number 100, but that 100 wasn't like 103. People thought a toy train was a lot like a real train but that a real train was not like a toy train. People often thought that a son resembled his father, but if you asked them if the father resembled his son, they just looked at you strangely. "The directionality and asymmetry of similarity relations are particularly noticeable in similes and metaphors," Amos wrote. "We say 'Turks fight like tigers' and not 'tigers fight like Turks.' Since the tiger is renowned for its fighting spirit, it is used as the referent rather than the subject of the simile. The poet writes 'my love is as deep as the ocean,' not 'the ocean is as deep as my love,' because the ocean epitomizes depth."

When people compared one thing to another — two people, two places, two numbers, two ideas — they did not pay much attention to symmetry. To Amos — and to no one else before Amos — it followed from this simple observation that all the theories that intellectuals had dreamed up to explain how people made similarity judgments had to be false. "Amos comes along and says you guys aren't asking the right question," says University of Michigan psychologist Rich Gonzalez. "What is distance? Distance is symmetric. New York to Los Angeles has to be the same distance as Los Angeles to New York. And Amos said, 'Okay, let's test that.' " If, on some mental map, New York sits a certain distance from Tel Aviv, Tel Aviv must sit precisely the same distance from New York. Yet you needed only to ask people to see that it did not: New York was not as much like Tel Aviv as Tel Aviv was like New York. "What Amos worked out was that whatever is going on is *not* a distance," says Gonzalez. "In one swoop he basically dismissed all theories that made use of distance. If you have a distance concept in your theory you are automatically wrong."

Amos had his own theory, which he called "features of similarity."* He argued that when people compared two things, and judged their similarity, they were essentially

* A paper by this name did not appear until 1977, but it grew from ideas he'd formed a decade earlier as a graduate student.

making a list of features. These features are simply *what they notice* about the objects. They count up the noticeable features shared by two objects: The more they share, the more similar they are; the more they don't share, the more dissimilar they are. Not all objects have the same number of noticeable features: New York City had more of them than Tel Aviv, for instance. Amos built a mathematical model to describe what he meant — and to invite others to test his theory, and prove him wrong.

Many have tried. Before he traveled to Stanford in the 1980s to study for his doctorate with Amos, Rich Gonzalez had read "Features of Similarity" several times. Upon arrival, he found his way to Amos's office, introduced himself, and asked what he thought was a killer question: "What about a three-legged dog?" Two three-legged dogs are obviously more similar to each other than a three-legged dog is to a four-legged dog. Yet a three-legged dog shares exactly the same number of features with a four-legged dog as it does with a three-legged dog. Ergo, an exception to Amos's theory! "I went in thinking, 'I'm outsmarting Amos,' " recalls Gonzalez. "He just looked at me like, *Really? That's the best you can come up with?* I think there might have been an initial glare, but then he was nice about it — and he said, 'The absence of a feature is a feature.' " Amos had written that into his original paper. "Similarity increases with the addition of common features and/or deletion of distinctive features."

From Amos's theory about the way people made judgments of similarity spilled all sorts of interesting insights. If the mind, when it compares two things, essentially counts up the features it notices in each of them, it might also judge those things to be at once more similar and more dissimilar to each other than some other pair of things. They might have both a lot in common and a lot *not* in common. Love and hate, and funny and sad, and serious and silly: Suddenly they could be seen — as they feel — as having more fluid relationships to each other. They weren't simply opposites on a fixed mental continuum; they could be thought of as similar in some of their features and different in others. Amos's theory also offered a fresh view into what might be happening when people violated transitivity and thus made seemingly irrational choices.

When people picked coffee over tea, and tea over hot chocolate, and then turned around and picked hot chocolate over coffee — they weren't comparing two drinks in some holistic manner. Hot drinks didn't exist as points on some mental map at fixed distances from some ideal. They were collections of features. Those features might become more or less noticeable; their prominence in the mind depended on the context in which they were perceived. And the choice created its own context: Different features might assume greater prominence in the mind when the coffee was being compared to tea (caffeine) than when it was being compared to hot chocolate (sugar). And what was true of drinks might also be true of people, and ideas, and emotions.

The idea was interesting: When people make decisions, they are also making judgments about similarity, between some object in the real world and what they ideally want. They make these judgments by, in effect, counting up the features they notice. And as the noticeability of features can be manipulated by the way they are highlighted, the sense of how similar two things are might also be manipulated. For instance, if you wanted two people to think of themselves as more similar to each other than they otherwise might, you might put them in a context that stressed the features they shared. Two American college students in the United States might look at each other and see a total stranger; the same two college students on their junior year abroad in Togo might find that they are surprisingly similar: They're both Americans!

By changing the context in which two things are compared, you submerge certain features and force others to the surface. "It is generally assumed that classifications are determined by similarities among the objects," wrote Amos, before offering up an opposing view: that "the similarity of objects is modified by the manner in which they are classified. Thus, similarity has two faces: causal and derivative. It serves as a basis for the classification of objects, but is also influenced by the adopted classification." A banana and an apple seem more similar than they otherwise would because we've agreed to call them both fruit. Things are grouped together for a reason, but, once they are grouped, their grouping causes them to seem more like each other than they otherwise would. That is, the mere act of classification reinforces stereotypes. If you want to weaken some stereotype, eliminate the classification.

Amos's theory didn't exactly contribute to the existing conversation about how people made judgments of similarity. It took over the entire conversation. Everyone else at the party just circled around Amos and listened. "Amos's approach to doing science wasn't incremental," said Rich Gonzalez. "It proceeded by leaps and bounds. You find a paradigm that is out there. You find a general proposition of that paradigm. And you destroy it. He saw himself doing a negative style of science. He used the word a lot: *negative.* This turns out to be a very powerful way of doing social science." That's how Amos would begin: by undoing the mistakes of others. As it turned out, other people had made some other mistakes.

4
Errors

By the time Amos returned to Israel in the fall of 1966, he'd been gone for five years. His oldest friends naturally compared the returning Amos to the Amos of their memories. They noticed a couple of changes. The Amos who returned from America appeared to them more serious about his work, and to have acquired a whiff of professionalism. He was now an assistant professor, with his own office at Hebrew University. He kept it famously spare. There was never anything on his desk but a mechanical pencil, and, if Amos was seated at it, an eraser and the crisply ordered file of whatever project he happened to be working on. When he'd left for the United States, he hadn't owned a suit. When he showed up at Hebrew University in a light blue suit, people were genuinely shocked, and not just by the color. "This was inconceivable," says Avishai Margalit. "This was something you didn't do. A tie was the symbol of the bourgeoisie. I remember the first time I saw my father in a suit and a tie. It was like finding your father with a whore." Otherwise Amos was unchanged: the last to go to bed at night, the life of every party, the light to which all butterflies flew, and the freest, happiest, and most interesting person anyone knew. He still did only what he wanted to do. Even his new interest in wearing a suit was more peculiarly Amos than it was bourgeois. Amos chose his suits only by the number and size of the jacket pockets. Along with an interest in pockets, he had what amounted to a fetish for briefcases, and acquired dozens of them. He'd returned from five years in the most materialistic culture on the face of the earth with a desire only for objects that might help him impose order on the world around him.

Along with a new suit, Amos also had a wife. In Michigan, three years earlier, he had met a fellow psychology student named Barbara Gans. They'd started dating after a year. "He told me he didn't want to go back to Israel alone," said Barbara. "And so we got married." She'd grown up in the Midwest and had never been out of the United States. What Europeans often said about Americans — how wildly informal and improvisational they were — was, to her, even more true of Israelis. "All you had were rubber bands and masking tape, so you fixed things with rubber bands and masking tape," she said. Though materially poor, Israel felt to her rich in other ways. Israelis — at least the Jewish ones — seemed all to earn roughly the same amount of money, and to have their basic needs met.

There weren't many luxuries. She and Amos had no phone and no car, but neither did most of the people they knew. The shops were all small and particular. There was the knife sharpener and the stonecutter and the falafel seller. If you needed a carpenter or a painter you didn't bother to phone them, even if you owned a phone, because they never answered. You went downtown in the afternoon and hoped to bump into them. "Everything was personal, all transactions. The standard joke was: Someone runs out of their burning house to ask a friend on the street if they know someone in the Fire Department." There was no television, but there were radios everywhere, and when the BBC came on everyone stopped whatever they were doing to listen. Those words felt consistently urgent. "Everyone was on alert," said Barbara. The tension in the air wasn't at all like the strife in the United

States over the Vietnam War. In Israel the danger felt present and personal: If the Arabs at every border ever stopped fighting among themselves, there was a sense, Barbara said, that they could overrun the country in a matter of hours and kill *you.*

The students at Hebrew University, where Barbara was given a psychology class to teach, seemed to be intent mainly on catching their professors in error. They were shockingly aggressive and lacking in deference. One student had so insulted a visiting American intellectual by interrupting his talk with derisive comments that university officials demanded he seek out the American and apologize. "I'm sorry if I have hurt your feelings," the student had said to the visiting dignitary, "but, you see, the talk was *so* bad!" For the final exam in one psychology class, the undergraduates were handed a published piece of research and told to find the flaw in it. On Barbara's second day, ten minutes into her lecture, a student in the back of the room screamed out, "Not true!" and no one seemed to think anything of it. A distinguished Hebrew University professor delivered a paper titled "What Is Not What in Statistics," after which a student in the audience announced, loudly enough for many to hear, "This will guarantee him a place in *Who Is Not Who in Statistics*!"

And yet at the same time, Israel took its professors more seriously than America did. Israeli intellectuals were presumed to have some possible relevance to the survival of the Jewish state, and the intellectuals responded by at least pretending to be relevant. In Michigan, Barbara and Amos had lived entirely within the university and spent their time with other academic types. Here they mixed with politicians and generals and journalists and others involved directly in running the country. In his first few months back, Amos gave talks about the latest decision-making theories to the generals in the Israeli army and the Israeli Air Force — even though the practical application of the theories was, to put it mildly, unclear. "I've never seen a country so concerned with keeping its officials abreast on new developments in academics," Barbara wrote to her family back home in Michigan.

And of course everyone was in the army, even the professors, and so it was impossible even for the most rarefied intellectual to insulate himself from the risks facing the entire society. All were exposed equally to the whims of dictators. That truth was hammered home to Barbara six months after she arrived, on May 22, 1967, when Egyptian president Gamal Abdel Nasser announced that he was closing the Straits of Tiran to Israeli ships. Most Israeli trade passed through the straits, and the announcement was taken as an act of war. "Amos came home one day and said, *The army is going to come for me.*" He rooted around and found a trunk that held his old paratrooper's uniform. It still fit him. At ten o'clock that night the army came for him.

It had been five years since Amos last jumped from an airplane; he was given an infantry unit to command. The entire country prepared for war — and at the same time tried to judge what kind of war it would be. In Jerusalem, those who remembered the war of independence feared another siege and emptied the stores of canned goods. People found it hard to assign probabilities to the potential outcomes: A war with Egypt alone would probably be ugly but survivable; a war with the combined Arab states might mean total annihilation. The Israeli government arranged quietly for the public parks to be consecrated, to allow them to be used as mass graves. The entire country mobilized. Private cars took over the bus routes — as all the buses had been taken by the army. Schoolchildren delivered the milk and the mail. Israeli Arabs, who weren't allowed to serve in the army, volunteered for the jobs left by Jewish conscripts. All the while an apocalyptic wind blew in from the desert. The sensation was like nothing Barbara had ever experienced. No matter how much you drank you felt thirsty; no matter how wet the laundry, it was dry inside of thirty minutes. It was 95 degrees, but standing in the desert gale you hardly noticed it was hot. She went to a kibbutz on the border just outside Jerusalem to help dig trenches. The man in his forties in charge of the volunteers had lost his leg in the war of independence and wore a prosthetic. He was a poet. He hobbled about, and worked on a poem.

Before the fighting began, Amos came home twice. Barbara was struck by how casually her new husband tossed his Uzi on the bed before taking a shower. No big deal! The country was in a state of panic, but Amos seemed unconcerned. "He told me, 'There is no reason to worry. It will depend upon airpower, and we have it. Our Air Force will destroy their planes.' " On the morning of June 5, with Egypt's army massed along the Israeli border, the Israeli Air Force launched a surprise attack. In a few hours Israeli pilots destroyed four hundred or so planes — virtually the entire Egyptian Air Force. Then the Israeli army rolled into the Sinai. By June 7 Israel was at war on three fronts against the armies of Egypt, Jordan, and Syria. Barbara went to a bomb shelter in Jerusalem and passed the time sewing sandbags.

It was later reported that, before the war, President Nasser had spoken with Ahmad Shukairy, the founder of the recently formed Palestine Liberation Organization. Nasser had proposed that Jews who survived the war be returned to their home countries; Shukairy had replied that there was no need to worry about it, as there wouldn't be any Jewish survivors. The war started on a Monday. The following Saturday the radio announced that it was over. Israel had won such a one-sided victory that it felt to many Jews less like a modern-day war than a miracle from the Bible. The country was suddenly more than twice as big as it had been a few days earlier, and controlled the Old City of Jerusalem, along with all the holy places. Just a week before, it had been the size of New Jersey; now it was bigger than Texas, with far more defensible borders. The radio stopped airing battle reports and played joyous Hebrew songs about Jerusalem. Here was another way Israel was different from the United States: Its wars were short, and someone always won.

On Thursday Barbara got a message from a soldier in Amos's unit; he let her know that Amos was alive. On Friday Amos drove up to their desert-beige apartment building in an army jeep and told her to hop in. Together they drove around the newly conquered West Bank. Along the way were strange and wonderful sights: warm reunions in the Old City of Jerusalem between Arab and Jewish shopkeepers, separated since 1948. A line of Arab men walking arm in arm up Ruppin Boulevard, in the Jewish Quarter, and pausing at the stoplights to clap . . . for the stoplights. The West Bank they found littered with burned-out Jordanian tanks and jeeps and empty tuna fish cans left by Israelis who had already come to picnic. They ended up in East Jerusalem, at the half-built summer palace of Jordan's King Hussein, where Amos was now stationed, along with a couple of hundred other Israeli soldiers. "That villa was really a shock," Barbara wrote to her family in Michigan that night, "combining the worst of Arabic taste with the worst of Miami Beach."

Later came the funerals. "This morning the figures were published in the newspaper

— 679 dead, 2563 wounded," Barbara wrote in a letter home. "Though the numbers are small, so is the country, so everyone can count the dead among his friends." Amos had lost one of his men in an attack that he had led on a monastery on top of a hill in Bethlehem. Elsewhere on the battlefield, one of his best friends from childhood had been killed by a sniper, and several Hebrew University professors had been killed or wounded. "I grew up in the Vietnam War and I hadn't known anyone who had gone to Vietnam, much less died there," said Barbara. "I knew four people who were killed in the Six-Day War — and I'd only been there six months."

For a week or so after the war, Amos camped at King Hussein's summer palace. He was then installed briefly as military governor of Jericho. Hebrew University was turned into a prisoner-of-war camp. But classes at the university started again on June 26, and the professors who had fought in the war were expected to resume their former posts without a lot of fuss. Among them was Amnon Rapoport, who had returned with Amos to Israel, joined him in Hebrew University's Department of Psychology, and taken his natural place as Amos's closest friend. When Amos set off with his infantry unit, Amnon had climbed into another tank and rolled back into Jordan. His tanks had taken the lead in breaking through the Jordanian army's front lines. This time Amnon had to admit to himself that this business of leaping into and out of wars had left him in a less than tranquil state of mind. "I mean, how is it possible? I am a young assistant professor. And they take me and within twenty-four hours I start killing people and become a killing machine. I didn't know how to put it together. The dreams troubled me for several months. Amos and I talked about it: how to reconcile these two sides of life. Professor and killer."

He and Amos had always assumed that they would work jointly to explore how people made decisions, but Amos was attached at the hip to Israel, and Amnon, once again, just wanted to get away. The problem, to Amnon, wasn't just the constant warfare. The idea of working with Amos had lost its allure. "He was so dominating, intellectually," said Amnon. "I realized that I didn't want to stay in the shadow of Amos all my life." In 1968 Amnon took off for the United States, became a professor at the University of North Carolina, and left Amos without anyone to talk to.

■ ■ ■ ■

In early 1967 Avishai Henik was twenty-one years old and working on a kibbutz in range of the Golan Heights. Every now and then the Syrians above him fired shells down on the kibbutz, but Avi didn't give it much thought. He'd just finished his army service and, even though he had been a poor student in high school, was thinking of going to university. In May 1967 he was trying, without a great deal of success, to decide what he would study, and the Israeli army called him back into service. If they were calling him, Avi assumed, there was going to be a war. He joined a unit of maybe one hundred and fifty paratroopers, most of whom he'd never laid eyes on.

Ten days later the war broke out. Avi had never seen combat. At first his commanding officers said that he was going to parachute into the Sinai and fight Egyptians. Then they changed their minds and ordered Avi's unit to board buses for Jerusalem, where a second front, with Jordan, had opened. In Jerusalem, there were two points of attack on the Jordanian troops entrenched just outside the Old City. Avi's unit slipped through the Jordanian front lines without firing a shot. "The Jordanians didn't even notice," he said. Hours later, a second Israeli paratrooper unit followed and was cut to bits: Avi's unit had gotten lucky. Once past the front lines, his unit approached the old walls. "That's when the shooting started," he said. Avi found himself trotting right beside a young man he liked named Moishe — Avi had only just met him a few days earlier, but he'd remember his face forever. A bullet struck Moishe and he fell. "He was dead in a minute." Avi moved on with the sense that at any moment he might die, too. "I was terrified," he said. "Really afraid." His unit fought their way through the Old City, and along the way ten more men were killed. "It was one here, one there." Avi recalled images and dramatic moments: Moishe's face; the Jordanian mayor of Jerusalem approaching his unit waving a white flag, standing beside the Wailing Wall. The last was incredible. "I was shocked. I'd seen it in pictures. And now I am standing right beside it." He turned to his commander and said how happy he was, and his commander replied, "Well, Avishai, you will not be happy tomorrow when you hear how many have been killed." Avi found a phone and called his mother and said simply, "I'm alive."

Avi's Six-Day War wasn't over. Having taken the Old City of Jerusalem, the surviving paratroopers in his unit were dispatched to the Golan Heights: Now they would fight Syrians. Along the way they met a middle-aged woman who came up to them and said, "You are paratroopers — has anyone seen my Moishe?" None of them had the courage to tell her what had happened to her son. Once they walked into the shadow of the Golan Heights, they were told their assignment: They would ascend in helicopters, jump out, and attack the Syrian troops in their trenches. Hearing this, Avi became oddly but completely certain that he was about to die. "I had the feeling that if I didn't die in Jerusalem, I would die in the Golan Heights," he said. "You don't get two chances." His commanding officer assigned him to walk point in the Syrian trenches — he would run in the front of a line of Israeli paratroopers until he was either killed or out of bullets.

Then — the very morning they were to go — the Israeli government announced that there would be a cease-fire at 6:30 p.m. For a brief moment Avi felt as if his life had been handed back to him. And yet his commanding officer insisted on proceeding with the attack. Avi couldn't understand it and summoned the nerve to ask his commanding officer why. Why go when the war will be over in a few hours? "He said, 'Avi, you are so naive. Do you think we will not take the Golan Heights even though there will be a cease-fire?' I said, 'Okay, prepare to die.' " With Avi in the lead, the paratrooper battalion stormed the Golan Heights in helicopters and leapt into the Syrian trenches. And the Syrians were gone. The trenches were empty.

After the war Avi, by then twenty-two years old, finally decided what he would study: psychology. Had you asked him just then why he picked psychology, "I would say I want to understand the human soul. Not the mind. The soul." Hebrew University had no room for him, so he went to a new university south of Tel Aviv called the University of the Negev. The campus was in Beersheba. He took two classes from a professor named Danny Kahneman, who was moonlighting because his job at Hebrew University didn't pay enough. The first was an introduction to statistics, which sounded deadly, only it wasn't. "He made it real by taking all these examples from life," recalled Avi. "He wasn't just teaching statistics. He was teaching: what is the meaning of all this?"

Danny was then helping the Israeli Air Force to train fighter pilots. He'd noticed that the instructors believed that, in teaching men to fly jets, criticism was more useful than praise. They'd explained to Danny that he only needed to see what happened after they praised a pilot for having performed especially well, or criticized him for performing especially badly. The pilot who was praised always performed worse the next time out, and the pilot who was criticized always performed better. Danny watched for a bit and then explained to them what was actually going on: The pilot who was praised because he had flown exceptionally well, like the pilot who was chastised after he had flown exceptionally badly, simply were regressing to the mean. They'd have tended to perform better (or worse) even if the teacher had said nothing at all. An illusion of the mind tricked teachers — and probably many others — into thinking that their words were less effective when they gave pleasure than when they gave pain. Statistics wasn't just boring numbers; it contained ideas that allowed you to glimpse deep truths about human life. "Because we tend to reward others when they do well and punish them when they do badly, and because there is regression to the mean," Danny later wrote, "it is part of the human condition that we are statistically punished for rewarding others and rewarded for punishing them."

The other class Danny taught was about perception: how the senses interpreted and, occasionally, misled. "Let me tell you: After two classes it was clear that this guy was brilliant," said Avi. Danny recited long passages from the Talmud in which the rabbis described day turning to night, and night turning to day, then asked the class: What colors are these rabbis seeing at that moment, when day turns to night? What did psychology have to say about the way the rabbis saw the world around them? Then he told them about the Purkinje effect — named for the Czech physiologist who had first described it, in the early nineteenth century. Purkinje had noticed that colors that appeared brightest to the human eye in broad daylight appeared the darkest at dusk. And so, for instance, what the rabbis saw as vividly red in the morning might appear, in contrast to other colors, almost colorless in the evening. Danny seemed to have in his head not only every strange phenomenon ever uncovered by anyone but an ability to describe them all in ways that led a student to see the world just a bit differently. "And he came to class with nothing!" said Avi. "He just came in and started talking."

A part of Avi couldn't quite believe the spontaneity of Danny's performances. He wondered if perhaps Danny had memorized his lectures and was just showing off." That suspicion was dispelled the day that Danny arrived to class and asked for help. "He came to me," recalled Avi, "and he said, 'Avi, my students at Hebrew University want me to give them something in writing, and I don't have anything. I saw you writing notes. Can I have them so I have something to give them?' . . . Everything was in his head!"

Avi soon learned that Danny expected his students to stuff their minds in much the same way that he had. Toward the end of his class on perception, Avi was called to army reserve duty. He went to Danny to say that, sadly, he needed to leave to patrol some remote border, and so he didn't see how he could keep up with the work and had to drop out of the class. "Danny said to me, 'It's okay, just learn the books.' And I said, 'What do you mean, just learn the books?' And he said, 'Take the books with you and *memorize* them.' " And so that's what Avi had done. He returned to Danny's classroom just in time for the final exam.

He'd memorized the books. Before Danny handed back the exams to the students, he asked Avi to raise his hand. "I raised my hand — what did I do this time? Danny says, 'You got 100 percent. And if someone gets a grade like this it should be said publicly.' "

After studying with this moonlighting professor from Hebrew University, Avi made two decisions: He would himself become a psychologist. And he would study at Hebrew University. He assumed that Hebrew University must be a magical place where the professors were geniuses who inspired their students to new heights of passion for their subjects. And so for graduate school Avi went to Hebrew University. At the end of his first year, the head of Hebrew University's Department of Psychology, surveying students, pulled Avi aside. *How are your teachers?* he asked.

They're okay, said Avi.

Okay? said the department head. *Just okay? Why are they only okay?*

I had this one teacher in Beersheba . . . , Avi started to say.

The department head immediately sensed what had happened. *Oh,* he said, *You're comparing them to Danny Kahneman. You can't do that. It's not fair to them. There's a category of teacher called Kahnemans. You cannot compare teachers to Kahnemans. You can say this guy is bad or good compared to others. That's okay. But not to Kahneman.*

Inside the classroom Danny was simply a bold genius. Outside the classroom — well, Avi was surprised by the volatility of Danny's state of mind. One day on campus, he ran into Danny and found him in a seriously dark mood — unlike anything Avi had ever seen. A student had just given him a bad review, Danny explained, and he thought that maybe he was all washed up. "He even asked me, 'I'm still the same man, right?' " It was obvious to Avi, and to everyone else but Danny, that the student was a fool. "Danny was the best teacher at Hebrew University," said Avi, "but it was very hard to convince him that the review didn't matter — that he was excellent." This was just the first of many sources of complication for Danny Kahneman: He was unusually inclined to believe the worst anyone said about him. "He was very insecure," Avi said. "This is part of his character."

To those he saw every day, Danny seemed unknowable. The picture people had in their minds of him was ever-shifting, like one of those sketches used for experiments by the

Gestalt psychologists. "He was moody in the extreme," said a former faculty colleague. "You never knew which Danny you were going to meet. He was very vulnerable. Starving for admiration and affection. Very edgy. Very impressionable. But could get easily insulted." He smoked two packs of cigarettes a day. He'd married, and his wife had given birth to a son and a daughter, but Danny still seemed to others to live entirely through his work. "He was very much task-oriented," said Zur Shapira, a student of Danny's who later became a professor at New York University. "You would not say he was a happy person." His moods put distance between Danny and other people, a bit like the distance caused by intense grief. "Women felt the urge to care for him," says Yaffa Singer, who worked with Danny in the Israeli army's psychology unit. "He was always in doubt," said Dalia Etzion, who served as Danny's teaching assistant. "I remember coming to him and he was blue. He was teaching, and he said, 'I'm sure the students don't like me.' I thought: What does it matter? And it was bizarre. Because the students love him." Another colleague said, "He was like Woody Allen, without the humor."

Danny's volatility was a weakness and, less

obviously, also a strength. It led him, almost inadvertently, to broaden himself. It turned out that Danny never really had to decide what kind of psychologist he would be. He could be, and would be, many different kinds of psychologists. At the same time that he was losing his faith in his ability to study personality, he was building a laboratory in which he might study vision. Danny's lab had this bench where subjects would be immobilized in a device constructed for that purpose, with their mouths stuck in an impression of their own teeth, while Danny flashed various signals at their pupils. The only way to understand a mechanism such as the eye, he thought, was by studying the mistakes that it made. Error wasn't merely instructive; it was the key that might unlock the deep nature of the mechanism. "How do you understand memory?" he asked. "You don't study memory. You study forgetting."

In his vision lab, Danny searched for the ways people's eyes played tricks on them. When exposed to vanishingly brief flashes of light, for example, the brightness that the eye experienced wasn't some straightforward function of the brightness of the flash. It also depended on the length of the flash — was in fact a product of the length of the flash and its intensity. A one-millisecond flash with an intensity of 10X was indistinguishable from a ten-millisecond flash with an intensity of X. But when flashes of light were longer than about 300 milliseconds, the brightness looked the same to people, no matter how long the flash lasted. The point of bothering to discover this was unclear, even to Danny, except that there was demand for such stuff in psychology journals, and he thought that the measuring was itself good training for him. "I was doing science," he said. "And I was being very deliberate about what I was doing. I consciously viewed what I was doing as filling a gap in my education, something I needed to do to become a serious scientist."

This sort of science didn't come naturally to him. A vision lab demanded precision, and Danny was about as precise as a desert storm. In the chaos that was his office, his secretary got so tired of being asked to help him search for his scissors that she tied them by a string to his desk chair. Even his interests were chaotic: That the same person could be mentally following schoolkids into the wilderness to ask them how many people they wanted sleeping in their tent, and sticking grown-ups' teeth into a vise to study how their eyes worked, struck even other psychologists as odd. Personality testers were hunting for loose correlations between some trait and some behavior: tent choice and sociability, for example, or IQ and job performance. They didn't need to be precise, and they need know nothing about people as biological organisms. Danny's studies of the human eye felt less like psychology than ophthalmology.

He nursed along other interests, too. He wanted to study what was known to psychologists as "perceptual defense" but to everybody else as subliminal perception. (A wave of anxiety had swept the United States in the late 1950s, thanks to a book by Vance Packard, called *The Hidden Persuaders,* about the power of advertising to warp people's decisions by influencing them subconsciously. Peak craze came in New Jersey, where a market researcher claimed that he had spliced imperceptibly brief messages like "Hungry? Eat Popcorn!" and "Drink Coca-Cola" into a movie and created a surge of demand for popcorn and Coke. He later confessed he'd made it all up.) Psychologists in the late 1940s had detected — or claimed to have detected — the mind's ability to defend itself from what it ostensibly did not want to perceive. When the experimenters flashed taboo words in front of subjects' eyes, for instance, the subjects read them as some less troubling word. At the same time, people were also influenced by the world around them in all sorts of ways without being entirely conscious of it: Stuff got into the mind without the mind's full awareness.

How did these unconscious processes work? How could a person understand a word well enough to distort it, without first having perceived it in some fashion? Was there perhaps more than one mechanism inside the mind at work? Did some part of the mind perceive incoming signals, say, while another part of the mind blocked them? "I was always interested in the question: 'Are there other ways to understand your experience?' " Danny said. "Perceptual defense was interesting because it seemed to get at unconscious life with proper experimental techniques." Danny designed some tests himself to see if, as he suspected, people were able to learn subconsciously. He showed subjects a series of playing cards or numbers, for example, and then asked them to predict what would come next. There was a hard-to-detect sequence in the cards or the numbers. If the subjects were able to sense the sequence, they would guess the next card or number more frequently than they would by chance — and they wouldn't know why! They'd have perceived the pattern without being aware of it. They'd have learned something subconsciously. Danny abandoned his experiments after he decided that his subjects had learned nothing.

That was another thing colleagues and students noticed about Danny: how quickly he moved on from his enthusiasms, how easily he accepted failure. It was as if he expected it. But he wasn't afraid of it. He'd try anything. He thought of himself as someone who enjoyed, more than most, changing his mind. "I get a sense of movement and discovery whenever I find a flaw in my thinking," he said. His theory of himself dovetailed neatly with his moodiness. In his darker moods, he became fatalistic — and so wasn't surprised or disturbed when he did fail. (He'd been proved right!) In his up moments he was so full of enthusiasm that he seemed to forget the possibility of failure, and would run with any new idea that came his way. "He could drive people up the wall with his volatility," said fellow Hebrew University psychologist Maya Bar-Hillel. "Something was genius one day and crap the next, and genius the next day and crap the next." What drove others crazy might have helped to keep Danny sane. His moods were grease for his idea factory.

If Danny's various intellectual pursuits had a common theme, other than his interest in them, it was hard for others to detect it. "He had no ability to see what is a waste of time and what is not," said Dalia Etzion. "He was willing to accept anything as possibly interesting." Suspicious of psychoanalysis ("I always thought it was a lot of mumbo jumbo"), he nevertheless accepted an invitation from the American psychoanalyst David Rapaport to spend a summer at the Austen Riggs Center in Stockbridge, Massachusetts. Each Friday morning the Austen Riggs psychoanalysts — some of the biggest names in the field — would gather to discuss a patient whom they had spent a month observing. All these experts would have by then written up their reports on the patient. After delivering their diagnoses, they would bring in the patient for an interview. One week Danny watched the psychoanalysts discuss a patient, a young woman. The night before they were meant to interview her, she committed suicide. None of the psychoanalysts — world experts who had spent a month studying the woman's mental state — had worried that she might kill herself. None of their reports so much as hinted at the risk of suicide. "Now they all agreed, how could we have missed it?" Danny recalled. "The signs were all there! It made so much sense to them after the fact. And so little sense before the fact." Any faint interest Danny might have had in psychoanalysis vanished. "I was aware at the time that this was very instructive," he said. Not about the troubled patients but about the psychoanalysts — or anyone else who was in a position to revise his forecast about the outcome of some uncertain event once he had knowledge of that outcome.

In 1965, he went to the University of Michigan for postdoctoral study with a psychologist named Gerald Blum. Blum was busy testing how powerful emotional states changed the way people handled various mental tasks. To do this he needed to induce in his subjects powerful emotional states. He did so with hypnosis. He'd first ask people to describe in detail some horrible life experience. He'd then give them a trigger to associate with the event — say, a card that read "A100." Then he'd hypnotize them, show them the card — and, sure enough, they'd instantly start to relive their horrible experience. Then he'd see how they performed some taxing mental task: say, repeating a string of numbers. “It was weird, and I did not take to it,” said Danny — though he did learn how to hypnotize people. “I ran some sessions with our best subject — a tall, thin guy whose eyes would bulge and his face redden as he was shown the A100 card that instructed him to have the worst emotional experience of his life for a few seconds.” Once again, it wasn’t long before Danny found himself undermining the validity of the entire enterprise. “One day I asked, ‘How about we give them a choice between *that* and a mild electric shock?’ ” he recalled. He figured that anyone given a choice between reliving the worst experience of his life and mild electric shock would choose the shock. None of the patients wanted the shock: They all said they’d much rather relive the worst experience of their lives. “Blum was horrified, because he wouldn’t hurt a fly,” said Danny. “And that’s when I realized that it was a stupid game. That it cannot be the worst experience of their lives. Somebody is faking. And so I got out of that field.”

That same year, a psychologist named Eckhard Hess wrote an article in *Scientific American* that caught Danny’s eye. (What didn’t?) Hess described the results of experiments he’d done measuring the dilation and constriction of the pupil in response to all sorts of stimuli. You showed a man the picture of a scantily dressed woman and his pupils expanded. The same thing happened when you showed a woman a picture of a good-looking man. On the other hand, if you showed people a picture of a shark, their pupils shrank. (Abstract art had the same effect, curiously.) If you gave people something tasty to drink, their pupils dilated; if you gave them something unpleasant (lemon juice or quinine), their pupils shrank. If you gave them tastes of five subtly different orange fizzy drinks, their pupils registered the degree of pleasure they got from each. People reacted incredibly quickly, before they were entirely conscious of which one they liked best. "The essential sensitivity of the pupil response," wrote Hess, "suggests that it can reveal preferences in some cases in which the actual taste differences are so slight that the subject cannot even articulate them."

The eye might offer a window into the mind. In Blum's hypnosis lab, with a psychologist named Jackson Beatty, whom he'd poached from Blum, Danny set out to investigate how the pupil responded when people were asked to perform various tasks that required mental effort: remember strings of digits, or distinguish sounds of different pitches. They were seeking to understand not whether the eye played tricks on the mind, but if the mind also played tricks on the eye. Or, as they put it, how "intense mental activity hinders perception." They found that it wasn't just emotional arousal that altered the size of the pupil: Mental effort had the same effect. There was, quite possibly, as they put it, "an antagonism between thinking and perceiving."

From Michigan, Danny planned to return to a tenured job at Hebrew University. When the university delayed its decision on whether to give him tenure, he refused to return. "I was very angry," he said. "I called and said, 'I'm not coming back.' " Instead, in the fall of 1966, he went to Harvard. (Three years at Berkeley had persuaded him that he was smart enough to play in the big leagues.) There he heard a talk, given by a young British psychologist named Anne Treisman, that sent him in yet another direction.

In the early 1960s, Treisman had picked up where the work of fellow Brits Colin Cherry and Donald Broadbent had left off. Cherry, a cognitive scientist, had identified what became known as the "cocktail party effect." The cocktail party effect was the ability of people to filter a lot of noise for the sounds they wished to hear — as they did when they listened to someone at a cocktail party. It was in those days a practical problem because of the design of air traffic control towers. In the early control towers, the voices of all the pilots who needed guidance were broadcast through loudspeakers. Air traffic controllers had to filter the voices to identify the relevant airplane. It was just assumed that they could ignore the voices that they needed to ignore in order to focus on the voice that required their attention.

Together with another British colleague, Neville Moray, Treisman set out to see just how selectively people listened when they listened selectively. "Nobody had done or was doing any research in the field of selective listening," she wrote in her memoir, "so we had it more or less to ourselves." She and Moray had put people in headphones attached to a two-channel tape recorder and piped two different passages of prose simultaneously into separate ears. Treisman asked the subjects to repeat back to her, as they listened, one of the passages. Afterward, she asked them what they had picked up from the passage they had supposedly ignored. It turned out that they hadn't entirely ignored it. Some words and phrases got through to the mind, even if they hadn't been invited. For instance, if their name was in the passage that they were assigned to ignore, people would often hear it.

This surprised Treisman, along with the few other people then paying attention to attention. "I thought at the time that attention was a complete filtering," said Treisman, "but it turns out that some kind of monitoring goes on. The question I had was, how do we do this? When, and how, does the content get through?" In her Harvard talk, Treisman proposed that people possessed, not an on-off switch that enabled them to pay attention to whatever they intended to pay attention to, but a more subtle mechanism that selectively weakened, rather than entirely blocked, background noise. That background noise might get through was, of course, not the happiest news for passengers in airplanes circling the control tower. But it was interesting.

Anne Treisman was on a flying visit to Harvard, where the demand to hear what she had to say was so great that her talk had to be moved to a big public lecture hall off campus. Danny left the talk filled with new enthusiasm. He asked to be deputized to look after Treisman and her traveling party — which included her mother, her husband, and their two small children. He gave them a tour of Harvard. "He was very eager to impress," said Treisman, "and so I let myself be impressed." It would be years before Danny and Anne left their marriages and married each other, but it took no time at all for Danny to engage Treisman's ideas.

In the fall of 1967 Danny had gotten over his feelings of being slighted and returned to Hebrew University, with the promise of tenure and an entirely new research program. It was now possible, with double-channel tape recorders, to measure how well people divided their attention, or switched their attention from one thing to another. It stood to reason that some people might be better at it than others, and that the ability might offer an advantage in certain lines of work. With this in mind Danny went to England, at the invitation of the Cambridge Applied Psychology Unit, to test professional soccer players. He thought that there might be a difference in the attention-switching abilities of players in the first (premier) league and players in the fourth league. He took the train from Cambridge to Arsenal — home to a premier league soccer team — with his heavy dual-track tape recorder beside him. He put the headphones on the players and tested their ability to switch from the message playing in one ear to the message playing in the other, and found . . . nothing. Or, at least, no obvious difference between them and the players in the lower-ranked league. A talent for playing soccer didn't require any special ability to switch attention.

"Then I thought, this could be critical in pilots," he recalled. He knew, from working with flight instructors, that the cadets training to fly fighter jets sometimes failed because they either couldn't divide their attention between tasks or were slow to pick up on seemingly unimportant but actually critical background signals. He returned to Israel and tested cadets who were training to fly jets for the Air Force. This time he found what he was looking for: The successful fighter pilots were better able to switch attention than the unsuccessful ones, and both were better at it than Israeli bus drivers. Eventually one of Danny's students discovered that you could predict, from how efficiently they switched channels, which Israeli bus drivers were more likely to have accidents.

There was a relentlessness in the way

Danny's mind moved from insight to application. Psychologists, especially the ones who became university professors, weren't exactly known for being useful. The demands of being an Israeli had forced Danny to find a talent in himself he might otherwise never have spotted. His high school friend Ariel Ginsberg thought that the Israeli army had made Danny more practical: The creation of a new interview system, and its effect on an entire army, had been intoxicating. The most popular class Danny taught at Hebrew University was a graduate seminar he called Applications of Psychology. Each week he brought in some real-world problem and told the students to use what they knew from psychology to address it. Some of the problems came from Danny's many attempts to make psychology useful to Israel. After terrorists started placing bombs in city trash cans — and one in the Hebrew University cafeteria in March 1969 that wounded twenty-nine students — Danny asked: What does psychology tell you that might be useful to the government, which is trying to minimize the public's panic? (Before they could arrive at an answer, the government removed the trash cans.)

Israelis in the 1960s lived with constant

change. Immigrants who had come from city life were channeled onto collective farms. The farms themselves underwent fairly constant technological upheaval. Danny designed a course to train the people who trained the farmers. "Reforms always create winners and losers," Danny explained, "and the losers will always fight harder than the winners." How did you get the losers to accept change? The prevailing strategy on the Israeli farms — which wasn't working very well — was to bully or argue with the people who needed to change. The psychologist Kurt Lewin had suggested persuasively that, rather than selling people on some change, you were better off identifying the reasons for their resistance, and addressing those. Imagine a plank held in place by a spring on either side of it, Danny told the students. How do you move it? Well, you can increase the force on one side of the plank. Or you can reduce the force on the other side. "In one case the overall tension is reduced," he said, "and in the other it is increased." And that was a sort of proof that there was an advantage in reducing the tension. "It's a key idea," said Danny. "Making it easy to change."

Danny was also training Air Force flight instructors to train fighter pilots. (But only on the ground: The one time they took him up in a plane he vomited into his oxygen mask.) How did you get fighter pilots to memorize a series of instructions? "We started making a long list," recalled Zur Shapira. "Danny says no. He tells us about 'The Magical Number Seven.' " "The Magical Number Seven, Plus or Minus Two: Some Limits on Our Capacity for Processing Information" was a paper, written by Harvard psychologist George Miller, which showed that people had the ability to hold in their short-term memory seven items, more or less. Any attempt to get them to hold more was futile. Miller half-jokingly suggested that the seven deadly sins, the seven seas, the seven days of the week, the seven primary colors, the seven wonders of the world, and several other famous sevens had their origins in this mental truth.

At any rate, the most effective way to teach people longer strings of information was to feed the information into their minds in smaller chunks. To this, Shapira recalled, Danny added his own twist. "He says you only tell them a few things — and get them to *sing* it." Danny loved the idea of the "action song." In his statistics classes he had actually asked his students to sing the formulas. "He forced you to engage with problems," said Baruch Fischhoff, a student who became a professor at Carnegie Mellon University, "even if they were complicated problems without simple solutions. He made you feel you could do something *useful* with this science."

A lot of the problems Danny threw at his students felt like pure whim. He asked them to design a currency so that it was hard to counterfeit. Was it better for bills of different denominations to resemble each other, as they did in the United States, thus leading anyone accepting them to examine them closely; or should they have a wide variety of colors and shapes so that they were harder to copy? He asked them how they would design a workplace to make it more efficient. (And of course they must be familiar with the psychological research showing that some wall colors led workers to be more productive than others.) Some of Danny's problems were so abstruse and strange that the student's first response was, *Um, we'll need to go to the library and get back to you on that.* "When we said that," recalled Zur Shapira, "Danny responded — mildly upset — by saying, 'You have completed a three-year program in psychology. You are by definition professionals. Don't hide behind research. Use your knowledge to come up with a plan.' "

But what were you supposed to say when Danny brought in a copy of a doctor's prescription from the twelfth century, sloppily written, in a language you didn't know a word of, and asked you to decode it? "Someone once said that education was knowing what to do when you don't know," said one of his students. "Danny took that idea and ran with it." One day Danny brought in a stack of those games in which the object is to guide a small metal ball through a wooden maze. The assignment he gave his students: Teach someone how to teach someone else how to play the game. "It would never occur to *anyone* that you could teach this," recalled one of the students. "The trick was to break it down into the component skills — learning how to hold your hand steady, learning how to tilt slightly to the right, and so on — then teach them separately and then, once you'd taught them all, put them together." The guy at the store who sold the games to Danny found the whole idea of it hysterical. But to Danny, useful advice, however obvious, was better than no advice at all. He asked his students to figure out what advice they would give to an Egyptologist who was having difficulty deciphering a hieroglyph. "He tells us that the guy is going slower and slower and getting more and more stuck," recalled Daniela Gordon, a student who became a researcher in the Israeli army. "Then Danny asks, 'What should he do?' No one could think of anything. And Danny says; 'He should take a nap!' "

Danny's students left every class with a sense that there was really no end to the problems in this world. Danny found problems where none seemed to exist; it was as if he structured the world around him so that it might be understood chiefly as a problem. To each new class the students arrived wondering what problem he might bring for them to solve. Then one day he brought them Amos Tversky.

5
The Collision

Danny and Amos had been at the University of Michigan at the same time for six months, but their paths seldom crossed; their minds, never. Danny had been in one building, studying people's pupils, and Amos had been in another, devising mathematical approaches to similarity, measurement, and decision making. "We had not had much to do with each other," said Danny. The dozen or so graduate students in Danny's seminar at Hebrew University were all surprised when, in the spring of 1969, Amos turned up. Danny never had guests: The seminar was his show. Amos was about as far removed from the real-world problems in Applications of Psychology as a psychologist could be. Plus, the two men didn't seem to mix. "It was the graduate students' perception that Danny and Amos had some sort of rivalry," said one of the students in the seminar. "They were clearly the stars of the department who somehow or other hadn't gotten in sync."

Before he left for North Carolina, Amnon Rapoport had felt that he and Amos disturbed Danny in some way that was hard to pin down. "We thought he was afraid of us or something," said Amnon. "Suspicious of us." For his part, Danny said he'd simply been curious about Amos Tversky. "I think I wanted a chance to know him better," he said.

Danny invited Amos to come to his seminar to talk about whatever he wanted to talk about. He was a little surprised that Amos didn't talk about his own work — but then Amos's work was so abstract and theoretical that he probably decided it had no place in the seminar. Those who stopped to think about it found it odd that Amos's work betrayed so little interest in the real world, when Amos was so intimately and endlessly engaged with that world, and how, conversely, Danny's work was consumed by real-world problems, even as he kept other people at a distance.

Amos was now what people referred to, a bit confusingly, as a "mathematical psychologist." Nonmathematical psychologists, like Danny, quietly viewed much of mathematical psychology as a series of pointless exercises conducted by people who were using their ability to do math as camouflage for how little of psychological interest they had to say. Mathematical psychologists, for their part, tended to view nonmathematical psychologists as simply too stupid to understand the importance of what they were saying. Amos was then at work with a team of mathematically gifted American academics on what would become a three-volume, molasses-dense, axiom-filled textbook called *Foundations of Measurement* — more than a thousand pages of arguments and proofs of how to measure stuff. On the one hand, it was a wildly impressive display of pure thought; on the other, the whole enterprise had a tree-fell-in-the-woods quality to it. How important could the sound it made be, if no one was able to hear it?

Instead of his own work, Amos talked to Danny's students about the cutting-edge research being done in Ward Edwards's lab at the University of Michigan. Edwards and his students were still engaged in what they considered to be an original line of inquiry. The specific study Amos described was about how people, in their decision making, responded to new information. As Amos told it, the psychologists had brought people in and presented them with two book bags filled with poker chips. Each bag contained both red poker chips and white poker chips. In one of the bags, 75 percent of the chips were white and 25 percent were red; in the other bag, 75 percent of the chips were red and 25 percent were white. The subject picked one of the bags at random and, without glancing inside the bag, began to pull chips out of it, one at a time. After extracting each chip, he'd give the psychologists his best guess of the odds that the bag he was holding was filled with mostly red, or mostly white, chips.

The beauty of the experiment was that there was a correct answer to the question: What is the probability that I am holding the bag of mostly red chips? It was provided by a statistical formula called Bayes's theorem (after Thomas Bayes, who, strangely, left the formula for others to discover in his papers after his death, in 1761). Bayes's rule allowed you to calculate the true odds, after each new chip was pulled from it, that the book bag in question was the one with majority white, or majority red, chips. Before any chips had been withdrawn, those odds were 50:50 — the bag in your hands was equally likely to be either majority red or majority white. But how did the odds shift after each new chip was revealed?

That depended, in a big way, on the so-called base rate: the percentage of red versus white chips in the bag. (These percentages were presumed to be known.) If you know that one bag contains 99 percent red chips and the other, 99 percent white chips, the color of the first chip drawn from the bag tells you a lot more than if you know that each bag contains only 51 percent red or white. But how much more does it tell you? Plug the base rate into Bayes's formula and you get an answer. In the case of two bags known to be 75 percent-25 percent majority red or white, the odds that you are holding the bag containing mostly red chips rise by three times every time you draw a red chip, and are divided by three every time you draw a white chip. If the first chip you draw is red, there is a 3:1 (or 75 percent) chance that the bag you are holding is majority red. If the second chip you draw is also red, the odds rise to 9:1, or 90 percent. If the third chip you draw is white, they fall back to 3:1. And so on.

The bigger the base rate — the known ratio of red to white chips — the faster the odds shift around. If the first three chips you draw are red, from a bag in which 75 percent of the chips are known to be either red or white, there's a 27:1, or slightly greater than 96 percent, chance you are holding the bag filled with mostly red chips.

The innocent subjects who pulled the poker chips out of the book bags weren't expected to know Bayes's rule. The experiment would have been ruined if they had. Their job was to guess the odds, so that the psychologists could compare those guesses with the correct answer. From their guesses, the psychologists hoped to get a sense of just how closely whatever was going on in people's minds resembled a statistical calculation when those minds were presented with new information. Were human beings good intuitive statisticians? When they didn't know the formula, did they still behave as if they did?

At the time, the experiments felt radical and exciting. In the minds of the psychologists, the results spoke to all sorts of real-world problems: How do investors respond to earnings reports, or patients to diagnoses, or political strategists to polls, or coaches to a new score? A woman in her twenties who receives from a single test a diagnosis of breast cancer is many times more likely to have been misdiagnosed than is a woman in her forties who receives the same diagnosis. (The base rates are different: Women in their twenties are far less likely to have breast cancer.) Does she sense her own odds? If so, how clearly? Life is filled with games of chance: How well do people play them? How accurately do they assess new information? How do people leap from evidence to a judgment about the state of the world? How aware are they of base rates? Do they allow what just happened to alter, accurately, their sense of the odds of what will happen next?

The broad answer to that last question coming from the University of Michigan, Amos reported to Danny's class, was that, yes, more or less, they do. Amos presented research done in Ward Edwards's lab that showed that when people draw a red chip from the bag, they do indeed judge the bag to be more likely to contain mostly red chips. If the first three chips they withdrew from a bag were red, for instance, they put the odds at 3:1 that the bag contained a majority of red chips. The true, Bayesian odds were 27:1. People shifted the odds in the right direction, in other words; they just didn't shift them dramatically enough. Ward Edwards had coined a phrase to describe how human beings responded to new information. They were "conservative Bayesians." That is, they behaved more or less as if they knew Bayes's rule. Of course, no one actually thought that Bayes's formula was grinding away in people's heads.

What Edwards, along with a lot of other social scientists, believed (and seemed to want to believe) was that people behaved as if they had Bayes's formula lodged in their minds. That view dovetailed with the story then winning the day in social science. It had been told best by the economist Milton Friedman. In a 1953 paper, Friedman wrote that a person shooting billiards does not calculate the angles on the table and the force imparted on the cue ball, and the reaction of one ball to another, in the way a physicist might. He just shot the ball in the right direction with roughly the right amount of force, as if he knew the physics. His mind arrived at more or less the right answer. How that happened didn't matter. Similarly, when a person calculates the odds of some situation, he does not do advanced statistics. He just behaves as if he does.

When Amos was done talking, Danny was baffled. *Was that it?* "Amos had described the research in the normal way that people describe research done by respected colleagues," said Danny. "You assume it is okay, and you trust the people who did it. When we look at a paper that has been published in a refereed journal, we tend to take it at face value — we assume that what the authors say must make sense — otherwise it would not have been published." And yet, to Danny, the experiment that Amos described sounded just incredibly stupid. After a person has pulled a red chip out of a bag, he is more likely than before to think the bag to be the one whose chips are mostly red: well, duh. What else is he going to think? Danny had had no exposure to the new research into the way people thought when they made decisions. "I had never thought much about thinking," he said. To the extent that Danny thought of thinking, he thought of it as *seeing* things. But this research into the human mind bore no relationship to what he knew about what people actually did in real life. The eye was often deceived, systematically. So was the ear.

The Gestalt psychologists he loved so much made entire careers out of fooling people with optical illusions: Even people who knew of the illusion remained fooled by it. Danny didn't see why thinking should be any more trustworthy. To see that people were not intuitive statisticians — that their minds did not naturally gravitate to the "right" answer — you needed only to sit in on any statistics class at Hebrew University. The students did not naturally internalize the importance of base rates, for instance. They were as likely to draw a big conclusion from a small sample as from a big sample. Danny himself — the best teacher of statistics at Hebrew University! — had figured out, long after the fact, that he had failed to replicate whatever it was that he had discovered about Israeli kids from their taste in tent sizes because he had relied on sample sizes that were too small. That is, he had tested too few kids to get an accurate picture of the population. He had assumed, in other words, that a few poker chips revealed the true contents of the book bag as clearly as a few big handfuls, and so he never fully determined what was in the bag.

In Danny's view, people were not conservative Bayesians. They were not statisticians of any kind. They often leapt from little information to big conclusions. The theory of the mind as some kind of statistician was of course just a metaphor. But the metaphor, to Danny, felt wrong. "I knew I was a lousy intuitive statistician," he said. "And I really didn't think I was stupider than anyone else."

The psychologists in Ward Edwards's lab were interesting to Danny in much the same way that the psychoanalysts at the Austen Riggs Center had been interesting to him after their patient had surprised them by killing herself. What interested him was their inability to face the evidence of their own folly. The experiment Amos had described was compelling only to someone already completely sold on the idea that people's intuitive judgment approximated the correct answer — that they were, at least roughly, good Bayesian statisticians.

Which was odd when you thought about it. Most real-life judgments did not offer probabilities as clean and knowable as the judgment of which book bag contained mostly red poker chips. The most you could hope to show with such experiments is that people were very poor intuitive statisticians — so poor they couldn't even pick the book bag that offered them the most favorable odds. People who proved to be expert book bag pickers might still stumble when faced with judgments in which the probabilities were far more difficult to know — say, whether some foreign dictator did, or did not, possess weapons of mass destruction. Danny thought, This is what happens when people become attached to a theory. They fit the evidence to the theory rather than the theory to the evidence. They cease to see what's right under their nose.

Everywhere one turned, one found idiocies that were commonly accepted as truths only because they were embedded in a theory to which the scientists had yoked their careers. "Just think about it," said Danny. "For decades psychologists thought that behavior is to be explained by learning, and they studied learning by looking at hungry rats learning to run to a goal box in a maze. That was the way it was done. Some people thought it was BS, but they were not smarter or more knowledgeable than the brilliant people who dedicated their career to what we now see as rubbish."

The people in this new field devoted to human decision making had become similarly blinded by their theory. *Conservative Bayesians.* The phrase was worse than meaningless. "It suggests people have the correct answer and they adulterate it — not any realistic psychological process that produces the judgments that people make," said Danny. "What do people actually *do* in judging these probabilities?" Amos was a psychologist and yet the experiment he had just described, with apparent approval, or at least not obvious skepticism, had in it no psychology at all. "It felt like a math exercise," said Danny. And so Danny did what every decent citizen of Hebrew University did when he heard something that sounded idiotic: He let Amos have it. "The phrase 'I pushed him into the wall' was often used, even for conversations among friends," explained Danny later. "The idea that everyone is entitled to his/her opinion was a California thing — that's not how we did things in Jerusalem."

By the end of the seminar, Danny must have sensed that Amos didn't particularly want to argue with him anymore. Danny went home and boasted to his wife, Irah, that he had won an argument with a brash younger colleague. Or anyway, that's how Irah remembered it. "This is, or was, an important aspect of Israeli discussions," Danny said. "They were competitive."

In the History of Amos there aren't a lot of examples of Amos losing an argument, and there are even fewer examples of Amos changing his mind. "You can never say he's wrong, even if he's wrong," said his former student Zur Shapira. It wasn't that Amos was rigid. In conversation he was freewheeling and fearless and open to new ideas — though perhaps more so if they did not openly conflict with his own. It was more that Amos had been right so often that, in any argument, "Amos is right" had become a useful assumption for all involved, Amos included. When asked for his memories of Amos, the first thing the Nobel Prize–winning Hebrew University economist Robert Aumann recalled was the one time he had surprised Amos with an idea. "I remember him saying, 'I didn't think of that,' " said Aumann. "And I remember it because there wasn't much Amos hadn't thought of."

Danny later suspected that Amos actually hadn't given much thought to the idea of the human mind as some kind of Bayesian statistician — the stuff with the book bags and poker chips wasn't his line of research. "Amos probably never had a serious discussion with anyone about that paper," said Danny. "And if he had, no one would have raised deep objections." People were Bayesian in the same way that people were mathematicians. Most people could work out that seven times eight equals fifty-six: so what if some could not? Whatever errors they made were random. It wasn't as if the human mind had some other way of doing math that led it to systematic error. If someone had asked Amos, "Do you think people are conservative Bayesians?," he might have said something like, "Certainly not every person, but as a description of the average person, it will do."

In the spring of 1969, at least, Amos wasn't overtly hostile to the reigning theories in social science. Unlike Danny, he wasn't dismissive of theory. Theories for Amos were like mental pockets or briefcases, places to put the ideas you wanted to keep. Until you could replace a theory with a better theory — a theory that better predicted what actually happened — you didn't chuck a theory out. Theories ordered knowledge, and allowed for better prediction. The best working theory in social science just then was that people were rational — or, at the very least, decent intuitive statisticians. They were good at interpreting new information, and at judging probabilities. They of course made mistakes, but their mistakes were a product of emotions, and the emotions were random, and so could be safely ignored.

But that day something shifted inside Amos. He left Danny's seminar in a state of mind unusual for him: doubt. After the seminar, he treated theories that he had more or less accepted as sound and plausible as objects of suspicion.

His closest friends, who found the change in him shocking, assumed that Amos had always had his doubts. For instance, on occasion he spoke of a problem experienced by Israeli army officers when they led troops through the desert. He'd experienced the problem himself. In the desert, the human eye had trouble judging shapes and distances. It was difficult to navigate. "That was something that really troubled Amos," said his friend Avishai Margalit. "In the army you had to navigate a lot. And he was very good at it. But it gave even him trouble. Traveling at night, you'd see a light in the distance: Was it close or far away? The water appeared as if it were a mile or less away — then it would take many hours to walk to it." The Israeli soldier couldn't protect his country if he didn't know the country, but the country was difficult to know. The army gave him maps, but the maps were often useless. A sudden storm could drastically alter the desert landscape; one day the valley was here, the next day it was over there. Leading soldiers in the desert, Amos had become sensitive to the power of optical illusion: An optical illusion could kill. Israeli army commanders in the 1950s and 1960s who became disoriented or lost their way also lost the obedience of their soldiers, as the soldiers understood that there was a short step from being lost to being dead. Amos wondered: If human beings had been shaped so carefully for their environment, why was their perception of that environment still prone to error?

There'd been other signs that Amos was less than wholly satisfied with the worldview of his fellow theorists in decision making. Just a few months before he'd spoken at Danny's seminar, for instance, he had been called back into the army, on reserve duty, and sent to the Golan Heights. There was no fighting to be done just then. His job was simply to command a unit in the newly acquired territory, gaze down upon Syrian soldiers, and judge from their movements if they were planning to attack. Under his command was Izzy Katznelson, who would go on to become a professor of mathematics at Stanford University. Like Amos, Katznelson had been a boy in Jerusalem during the 1948 war of independence; scenes from that year were seared into his memory. He remembered Jews running into the houses of Arabs who had fled and stealing whatever they could. "I thought, those Arabs are people like me: They didn't start the war and I didn't start the war," he said. He'd followed the noise inside one of the Arab houses and discovered yeshiva boys destroying the Arabs' grand piano — for the wood. Katznelson and Amos didn't talk about that; those were events best forgotten.

What they talked about was Amos's new curiosity about the way people judged the likelihood of uncertain events — for instance, the probability of an attack at that moment by the Syrian army. "We were standing looking at the Syrians," recalled Katznelson. "He was talking about probabilities, and how do you assign probabilities. He was interested in how, in 1956 [moments before the Sinai campaign], the government had made some estimates that there wouldn't be a war for five years, and other estimates that there wouldn't be a war for at least ten years. What Amos was pushing is that probability was not a given. People do not know how to do it properly."

If, since his return to Israel, there had indeed been a growing pressure along some fault line inside Amos's mind, the encounter with Danny had triggered the earthquake. Not long afterward, he bumped into Avishai Margalit. "I'm waiting in this corridor," said Margalit. "And Amos comes to me, agitated, really. He started by dragging me into a room. He said, *You won't believe what happened to me.* He tells me that he had given this talk and Danny had said, *Brilliant talk, but I don't believe a word of it.* Something was really bothering him, and so I pressed him. He said, 'It cannot be that judgment does not connect with perception. Thinking is not a separate act.' " The new studies being made about how people's minds worked when rendering dispassionate judgments had ignored what was known about how the mind worked when it was doing other things. "What happened to Amos was serious," said Danny. "He had a commitment to a view of the world in which Ward Edwards's research made sense, and that afternoon he saw the appeal of another worldview in which that research looked silly."

After the seminar, Amos and Danny had a few lunches together but then headed off in separate directions. That summer Amos left for the United States, and Danny for England, to continue his studies of attention. He had all these ideas about the possible usefulness of his new work on attention. In tank warfare, for instance. In his research, Danny was now taking people and piping one stream of digits into their left ear and another stream of digits into their right ear, and testing how quickly they could switch their attention from one ear to the other, and also how well they blocked their minds to sounds they were meant to be ignoring.

"In tank warfare, as in a Western shootout, the speed at which one can decide on a target and act on that decision makes the difference between life and death," said Danny later. He might use his test to identify which tank commanders could best orient their senses at high speed — who among them might most quickly detect the relevance of a signal, and focus his attention upon it, before he got blown to bits.

By the fall of 1969 Amos and Danny had both returned to Hebrew University. During their joint waking hours, they could usually be found together. Danny was a morning person, and so anyone who wanted him alone could find him before lunch. Anyone who wanted time with Amos could secure it late at night. In the intervening time, they might be glimpsed disappearing behind the closed door of a seminar room they had commandeered. From the other side of the door you could sometimes hear them hollering at each other, but the most frequent sound to emerge was laughter. Whatever they were talking about, people deduced, must be extremely funny. And yet whatever they were talking about also felt intensely private: Other people were distinctly not invited into their conversation. If you put your ear to the door, you could just make out that the conversation was occurring in both Hebrew and English. They went back and forth — Amos, especially, always switched back to Hebrew when he became emotional.

The students who once wondered why the two brightest stars of Hebrew University kept their distance from each other now wondered how two so radically different personalities could find common ground, much less become soul mates. "It was *very* difficult to imagine how this chemistry worked," said Ditsa Kaffrey, a graduate student in psychology who studied with them both. Danny was a Holocaust kid; Amos was a swaggering Sabra — the slang term for a native Israeli. Danny was always sure he was wrong. Amos was always sure he was right. Amos was the life of every party; Danny didn't go to the parties. Amos was loose and informal; even when he made a stab at informality, Danny felt as if he had descended from some formal place. With Amos you always just picked up where you left off, no matter how long it had been since you last saw him. With Danny there was always a sense you were starting over, even if you had been with him just yesterday. Amos was tone-deaf but would nevertheless sing Hebrew folk songs with great gusto. Danny was the sort of person who might be in possession of a lovely singing voice that he would never discover. Amos was a one-man wrecking ball for illogical arguments; when Danny heard an illogical argument, he asked, *What might that be true of?* Danny was a pessimist. Amos was not merely an optimist; Amos *willed* himself to be optimistic, because he had decided pessimism was stupid. *When you are a pessimist and the bad thing happens, you live it twice,* Amos liked to say. *Once when you worry about it, and the second time when it happens.* "They were very different people," said a fellow Hebrew University professor. "Danny was always eager to please. He was irritable and short-tempered, but he wanted to please. Amos couldn't understand why anyone would be eager to please. He understood courtesy, but eager to please — why??" Danny took everything so seriously; Amos turned much of life into a joke. When Hebrew University put Amos on its committee to evaluate all PhD candidates, Amos was appalled at what passed for a dissertation in the humanities. Instead of raising a formal objection, he merely said, "If this dissertation is good enough for its field, it's good enough for me. Provided the student can divide fractions!"

Beyond that, Amos was the most terrifying mind most people had ever encountered. "People were afraid to discuss ideas in front of him," said a friend — because they were afraid he would put his finger on the flaw that they had only dimly sensed. One of Amos's graduate students, Ruma Falk, said she was so afraid of what Amos would think of her driving that when she drove him home, in *her* car, she insisted that he drive. And now here he was spending all of his time with Danny, whose susceptibility to criticism was so extreme that a single remark from a misguided student sent him down a long, dark tunnel of self-doubt. It was as if you had dropped a white mouse into a cage with a python and come back later and found the mouse talking and the python curled in the corner, rapt.

But there was another story to be told, about now much Danny and Amos had in common. Both were grandsons of Eastern European rabbis, for a start. Both were explicitly interested in how people functioned when they were in a "normal" unemotional state. Both wanted to do science. Both wanted to search for simple, powerful truths. As complicated as Danny might have been, he still longed to do "the psychology of single questions," and as complicated as Amos's work might have seemed, his instinct was to cut through endless bullshit to the simple nub of any matter. Both men were blessed with shockingly fertile minds. And both were Jews, in Israel, who did not believe in God. And yet all anyone saw were their differences.

The most succinct physical manifestation of the deep difference between the two men was the state of their offices. "Danny's office was such a mess," recalled Daniela Gordon, who had become Danny's teaching assistant. "Scraps on which he'd scribbled a sentence or two. Paper everywhere. Books everywhere. Books opened to places he'd stopped reading. I once found my master's thesis open on page thirteen — I think that's where he stopped. And then you would walk down the hall three or four rooms, and you come to Amos's office . . . and there is nothing in it. A pencil on a desk. In Danny's office you couldn't find anything because it was such a mess. In Amos's office you couldn't find anything because there was nothing there." All around them people watched and wondered: Why were they getting along so well? "Danny was a high-maintenance person," said one colleague.

"Amos was the last one to put up with a high-maintenance person. And yet he was willing to go along. Which was amazing."

Danny and Amos didn't talk much about what they got up to when they were alone together, which just made everyone else more curious about what it was. In the beginning they were kicking around Danny's proposition — that people weren't Bayesians, or conservative Bayesians, or statisticians of any sort. Whatever human beings did when presented with a problem that had a statistically correct answer, it wasn't statistics. But how did you sell *that* to an audience of professional social scientists who were more or less blinded by theory? And how did you test it? They decided, in essence, to invent an unusual statistics test and give it to the scientists, and see how they performed. Their case would be built from evidence that consisted entirely of answers to questions they'd put to some audience — in this case, an audience of people trained in statistics and probability theory. Danny dreamed up most of the questions, many of which were sophisticated versions of the questions about red and white poker chips:

The mean IQ of the population of eighth

graders in a city is known to be 100. You have selected a random sample of 50 children for a study of educational achievement. The first child tested has an IQ of 150. What do you expect the mean IQ to be for the whole sample?

At the end of the summer of 1969, Amos took Danny's questions to the annual meeting of the American Psychological Association, in Washington, DC, and then on to a conference of mathematical psychologists. There he gave the test to roomfuls of people whose careers required fluency in statistics. Two of the test takers had written statistics textbooks. Amos then collected the completed tests and flew home with them to Jerusalem.

There he and Danny sat down to write together for the first time. Their offices were tiny, so they worked in a small seminar room. Amos didn't know how to type, and Danny didn't particularly want to, so they sat with notepads. They went over each sentence time and again and wrote, at most, a paragraph or two each day. "I had this sense of realization: Ah, this is not going to be the usual thing, this is going to be something else," said Danny. "Because it was *funny.*"

When Danny looked back on that time, what he recalled mainly was the laughter — what people outside heard from the seminar room. "I have the image of balancing precariously on the back legs of a chair and laughing so hard I nearly fell backwards." The laughter might have sounded a bit louder when the joke had come from Amos, but that was only because Amos had a habit of laughing at his own jokes. ("He was so funny that it was okay he was laughing at his own jokes.") In Amos's company Danny felt funny, too — and he'd never felt that way before. In Danny's company Amos, too, became a different person: uncritical. Or, at least, uncritical of whatever came from Danny. He didn't even poke fun in jest. He enabled Danny to feel, in a way he hadn't before, confident. Maybe for the first time in his life Danny was playing offense. "Amos did not write in a defensive crouch," he said. "There was something liberating about the arrogance — it was extremely rewarding to feel like Amos, smarter than almost everyone." The finished paper dripped with Amos's self-assurance, beginning with the title he had put on it: "Belief in the Law of Small Numbers." And yet the collaboration was so complete that neither of them felt comfortable taking the credit as the lead author; to decide whose name would appear first, they flipped a coin. Amos won.

"Belief in the Law of Small Numbers" teased out the implications of a single mental error that people commonly made — even when those people were trained statisticians. People mistook even a very small part of a thing for the whole. Even statisticians tended to leap to conclusions from inconclusively small amounts of evidence. They did this, Amos and Danny argued, because they believed — even if they did not acknowledge the belief — that any given sample of a large population was more representative of that population than it actually was.

The power of the belief could be seen in the way people thought of totally random patterns — like, say, those created by a flipped coin. People knew that a flipped coin was equally likely to come up heads as it was tails. But they also thought that the tendency for a coin flipped a great many times to land on heads half the time would express itself if it were flipped only a few times — an error known as "the gambler's fallacy." People seemed to believe that if a flipped coin landed on heads a few times in a row it was more likely, on the next flip, to land on tails — as if the coin itself could even things out. "Even the fairest coin, however, given the limitations of its memory and moral sense, cannot be as fair as the gambler expects it to be," they wrote. In an academic journal that line counted as a splendid joke.

They then went on to show that trained scientists — experimental psychologists — were prone to the same mental error. For instance, the psychologists who were asked to guess the mean IQ of the sample of kids, in which the first kid was found to have an IQ of 150, often guessed that it was 100, or the mean of the larger population of eight graders. They assumed that the kid with the high IQ was an outlier who would be offset by an outlier with an extremely low IQ — that every heads would be followed by a tails. But the correct answer — as produced by Bayes's theorem — was 101.

Even people trained in statistics and probability theory failed to intuit how much more variable a small sample could be than the general population — and that the smaller the sample, the lower the likelihood that it would mirror the broader population. They assumed that the sample would correct itself until it mirrored the population from which it was drawn. In very large populations, the law of large numbers did indeed guarantee this result. If you flipped a coin a thousand times, you were more likely to end up with heads or tails roughly half the time than if you flipped it ten times. For some reason human beings did not see it that way. "People's intuitions about random sampling appear to satisfy the law of small numbers, which asserts that the law of large numbers applies to small numbers as well," Danny and Amos wrote.

This failure of human intuition had all sorts of implications for how people moved through the world, and rendered judgments and made decisions, but Danny and Amos's paper — eventually published in the *Psychological Bulletin* — dwelled on its consequences for social science. Social science experiments usually involved taking some small sample from a large population and testing some theory on it. Say a psychologist thought that he had discovered a connection: Children who preferred to sleep alone on camping trips were somewhat less likely to participate in social activities than were children who preferred eight-person tents. The psychologist had tested a group of twenty kids, and they confirmed his hypothesis. Not every child who wanted to sleep alone was asocial, and not every child who longed for an eight-person tent was highly sociable — but the pattern existed. The psychologist, being a conscientious scientist, selects a second sample of kids — to see if he can replicate this finding. But because he has misjudged how large the sample needs to be if it is to stand a good chance of reflecting the entire population, he is at the mercy of luck.* Given the inherent variability of the small sample, the kids in his second sample might be unrepresentative, not at all like most children. And yet he treated them as if they had the power to confirm or refute his hypothesis.

The belief in the law of small numbers: Here was the intellectual error that Danny and Amos suspected that a lot of psychologists made, because Danny had made it. And Danny had a far better feel for statistics than most psychologists, or even most statisticians. The entire project, in other words, was rooted in Danny's doubts about

* A lot of psychologists at the time, including Danny, were using sample sizes of 40 subjects, which gave them only a 50 percent chance of accurately reflecting the population. To have a 90 percent chance of capturing the traits of the larger population, the sample size needed to be at least 130. To gather a larger sample of course required a lot more work, and thus slowed a research career.

his own work, and his willingness, which was almost an eagerness, to find error in that work. In their joint hands, Danny's tendency to look for his own mistakes became the most fantastic material. For it wasn't just Danny who made those mistakes: Everyone did. It wasn't just a personal problem; it was a glitch in human nature. At least that was their suspicion.

The test they administered to psychologists confirmed that suspicion. When seeking to determine if the bag they held contained mostly red chips, psychologists were inclined to draw, from very few chips, broad conclusions. In their search for scientific truth, they were relying far more than they knew on chance. What's more, because they had so much faith in the power of small samples, they tended to rationalize whatever they found in them.

The test Amos and Danny had created asked the psychologists how they would advise a student who was testing a psychological theory — say, that people with long noses are more likely to lie. What should the student do if his theory tests as true on one sample of humanity but as false on another? The question Danny and Amos put to the professional psychologists was multiple-choice. Three of the choices involved telling the student either to increase his sample size or, at the very least, to be more circumspect about his theory. Overwhelmingly, the psychologists had plunked for the fourth option, which read: "He should try to find an explanation for the differences between the two groups."

That is, he should seek to rationalize why in one group people with long noses are more likely to lie, while in the other they are not. The psychologists had so much faith in small samples that they assumed that whatever had been learned from either group must be generally true, even if one lesson seemed to contradict the other. The experimental psychologist "rarely attributes a deviation of results from expectations to sampling variability because he finds a causal 'explanation' for any discrepancy," wrote Danny and Amos. "Thus, he has little opportunity to recognize sampling variation in action. His belief in the law of small numbers, therefore, will forever remain intact."

To which Amos, by himself, appended: "Edwards . . . has argued that people fail to extract sufficient information or certainty from probabilistic data; he called this failure conservatism. Our respondents can hardly be described as conservative. Rather, in accord with the representation hypothesis, they tend to extract more certainty from the data than the data, in fact, contain." ("Ward Edwards was established," said Danny. "And we were taking pot shots — Amos was sticking his tongue out at him.")

By the time they were finished with the paper, in early 1970, they had lost any clear sense of their individual contributions. It was nearly impossible to say, of any given passage, whether more of some idea had come from Danny or from Amos. Far more easily determined, at least for Danny, was responsibility for the paper's confident, almost brazen, tone. Danny had always been a nervous scholar. "If I had written it alone, in addition to being tentative and having a hundred references, I would probably have confessed that I am only a recently reformed idiot," he said. "I could have done the paper all by myself. Except that if I had done it alone people would not have paid it attention. It had a star quality. And I attributed that quality to Amos."

He thought that their paper was funny and provocative and interesting and arrogant in a way he could never be on his own, but in truth he didn't think any more than that — and he didn't think Amos did, either. Then they gave the paper to a person they assumed would be a skeptical audience, a psychology professor at the University of Michigan named Dave Krantz. Krantz was a serious mathematician, and also one of Amos's coauthors on the impenetrable multivolume *Foundations of Measurement.* "I thought it was a stroke of genius," recalled Krantz. "I still think it is one of the most important papers that has ever been written. It was counter to all the work that was being done — which was governed by the idea that you were going to explain human judgment by correcting for some more or less minor error to the Bayesian model. It was exactly contrary to the ideas that *I* had. Statistics was the way you *should* think about probabilistic situations, but statistics was not the way people did it. Their subjects were all sophisticated in statistics — and even they got it wrong! Every question in the paper that the audience got wrong I felt the temptation to get wrong."

That verdict — that Danny and Amos's paper wasn't just fun but important — would eventually be echoed outside of psychology. "Over and over again economists say, 'If the evidence of the world tells you it is true, then people figure out what's true,' " says Matthew Rabin, a professor of economics at Harvard University. "That people are, in effect, very good statisticians. And if they aren't — well, they don't survive. And so if you are going down the list of things that are important in the world, the fact that people don't believe in statistics is pretty important."

Danny, being Danny, was slow to accept the compliment. ("When Dave Krantz said, 'It's a breakthrough,' I thought he was out of his mind.") Still, he and Amos were onto something far bigger than an argument about how to use statistics. The power of the pull of a small amount of evidence was such that even those who knew they should resist it succumbed. People's "intuitive expectations are governed by a consistent misperception of the world," Danny and Amos had written in their final paragraph. The misperception was rooted in the human mind. If the mind, when it was making probabilistic judgments about an uncertain world, was not an intuitive statistician, what was it? If it wasn't doing what the leading social scientists thought it did, and economic theory assumed that it did, what, exactly, was it doing?

6
The Mind's Rules

In 1960 Paul Hoffman, a professor of psychology at the University of Oregon with a special interest in human judgment, persuaded the National Science Foundation to give him sixty thousand dollars so that he could quit his teaching job and create what he described as a "center for basic research in the behavioral sciences." He'd never really enjoyed teaching all that much and was frustrated by how slowly academic life moved, especially in granting him promotions. And so he quit and bought a building in a leafy Eugene neighborhood that had most recently housed a Unitarian church, and renamed it the Oregon Research Institute. A private institution devoted exclusively to the study of human behavior, there was nothing in the world like it, and it soon attracted both curious assignments and unusual people. "Here brainy people, working in the proper atmosphere, go quietly about their task of finding out what makes us tick," a local Eugene paper reported.

The vagueness of that account became typical of descriptions of the Oregon Research Institute. No one really knew what the psychologists inside were up to — only that they could no longer say "I'm a professor" and leave it at that. After Paul Slovic left the University of Michigan to join Hoffman in his new research center, and his small children asked him what he did for a living, he would point to a poster that depicted a brain sectioned into its various compartments and say, "I study the mysteries of the mind."

Psychology had long been an intellectual garbage bin for problems and questions that for whatever reason were not welcome in other academic disciplines. The Oregon Research Institute became a practical extension of that bin. One early assignment came from a contracting company based in Eugene that had been hired to help build a pair of audacious skyscrapers in lower Manhattan, to be called the World Trade Center. The twin towers were to be 110 stories and built from light steel frames. The architect, Minoru Yamasaki, who had a fear of heights, had never designed any building higher than twenty-eight stories. The owner, New York Port Authority, planned to charge higher rents for the upper floors, and wanted the engineer, Les Robertson, to ensure that the high-paying tenants on the upper floors never sensed that the buildings moved with the wind. Realizing that this was not so much an engineering problem as a psychological one — how much could a building move before a person sitting at a desk on the ninety-ninth floor felt it? — Robertson turned to Paul Hoffman and the Oregon Research Institute.

Hoffman rented another building in another leafy Eugene neighborhood and built a room inside of it on top of the hydraulic wheels used to roll logs through Oregon's lumber mills. At the press of a button the entire room could be made to rock back and forth, silently, like the top of a Manhattan skyscraper in a breeze. All of this was done in secrecy. The Port Authority didn't want to alert its future tenants that they'd be swinging in the wind, and Hoffman worried that if his subjects knew they were in a building that moved, they would become more sensitive to movement and queer the experiment's results. "After they'd designed the room," recalled Paul Slovic, "the question was, how do we get people into the room without them knowing why?" And so after the "sway room" was built, Hoffman stuck a sign outside that read Oregon Research Institute Vision Research Center, and offered free eye exams to all comers. (He'd found a graduate student in psychology at the University of Oregon who happened also to be a certified optometrist.)

As the graduate student performed eye exams, Hoffman turned up the hydraulic rollers and made the room roll back and forth. The psychologists soon discovered that people in a building that was moving were far quicker to sense that something was off about the place than anyone, including the designers of the World Trade Center, had ever imagined. This is a strange room," said one. "I suppose it's because I don't have my glasses on. Is it rigged or something? It really feels funny." The psychologist who ran the eye exams went home every night seasick.*

* I owe some of this to a spectacular article about the construction and destruction of the World Trade Center towers by James Glanz and Eric Lipton, published in the *New York Times Magazine* a few days before the first anniversary of the attacks. William Poundstone's book *Priceless* offers a more detailed account of the sway room.

When they learned of Hoffman's findings, the World Trade Center's engineer, its architect, and assorted officials from the New York Port Authority flew to Eugene to experience the sway room themselves. They were incredulous. Robertson later recalled his reaction for the *New York Times:* "A billion dollars right down the tube." He returned to Manhattan and built his very own sway room, where he replicated Hoffman's findings. In the end, to stiffen the buildings, he devised, and installed in each of them, eleven thousand two-and-a-half-foot-long metal shock absorbers. The extra steel likely enabled the buildings to stand for as long as they did after they were struck by commercial airliners, and it allowed some of the fourteen thousand people who escaped to flee before the buildings collapsed.

For the Oregon Research Institute, the sway room was a bit of a diversion. Many of the psychologists who joined the place shared Paul Hoffman's interest in human judgment. They also shared an uncommon interest in Paul Meehl's book, *Clinical versus Statistical Prediction,* about the inability of psychologists to outperform algorithms when trying to diagnose, or predict the behavior of, their patients. It was the same book Danny Kahneman had read in the mid-1950s before he replaced the human judges of new Israeli soldiers with a crude algorithm. Meehl was himself a clinical psychologist, and kept insisting that of course psychologists like him and those he admired had many subtle insights that could never be captured by an algorithm. And yet by the early 1960s there was a swelling pile of studies that supported Meehl's initial pie-chucking skepticism of human judgment.*

If human judgment was somehow inferior to simple models, humanity had a big problem: Most fields in which experts

* In 1986, thirty-two years after the publication of his book, Meehl wrote an essay called "Causes and Effects of My Disturbing Little Book," in which he discussed the by then overwhelming evidence that expert judgment had its issues. "When you are pushing 90 investigations," wrote Meehl, "predicting everything from the outcome of football games to the diagnosis of liver disease[,] and when you can hardly come up with a half dozen studies showing even a weak tendency in favor of the clinician, it is time to draw a practical conclusion. . . . Not to argue ad hominem but to explain after the fact, I think this is just one more of the numerous examples of the ubiquity and recalcitrance of irrationality in the conduct of human affairs."

rendered judgments were not as data-rich, or as data-loving, as psychology. Most spheres of human activity lacked the data to build the algorithms that might replace the human judge. For most of the thorny problems in life, people would need to rely on the expert judgment of some human being: doctors, judges, investment advisors, government officials, admissions officers, movie studio executives, baseball scouts, personnel managers, and all the rest of the world's deciders of things. Hoffman, and the psychologists who joined his research institute, hoped to figure out exactly what experts were doing when they rendered judgments. "We didn't have a special vision," said Paul Slovic. "We just had a feeling this was important: how people took pieces of information and somehow processed that and came up with a decision or a judgment."

Interestingly, they didn't set out to explore just how poorly human experts performed when forced to compete with an algorithm. Rather, they set out to create a model of what experts were doing when they formed their judgments. Or, as Lew Goldberg, who had arrived in 1960 at the Oregon Research Institute by way of Stanford University, put it, "To be able to spot when and where human judgment is more likely to go wrong:

that was the idea." If they could figure out where the expert judgments were going wrong, they might close the gap between the expert and the algorithms. "I thought that if you understood how people made judgments and decisions, you could improve judgment and decision making," said Slovic. "You could make people better predictors and better deciders. We had that sense — though it was kind of fuzzy at the time."

To that end, in 1960, Hoffman had published a paper in which he set out to analyze how experts drew their conclusions. Of course you might simply ask the experts how they did it — but that was a highly subjective approach. People often said they were doing one thing when they were actually doing another. A better way to get at expert thinking, Hoffman argued, was to take the various inputs the experts used to make their decisions ("cues," he called these inputs) and infer from those decisions the weights they had placed on the various inputs. So, for example, if you wanted to know how the Yale admissions committee decided who got into Yale, you asked for the list of the information about Yale applicants that were taken into account — grade point average, board scores, athletic ability, alumni connections, type of high school attended, and so on. Then you watched the commitee decide, over and over, whom to admit. From the committee's many decisions you could distill the process its members had used to weigh the traits deemed relevant to the assessment of any applicant. You might even build a model of the interplay of those traits in the minds of the members of the committee, if your math skills were up to it. (The committee might place greater weight on the board scores of athletes from public schools, say, than on those of the legacy children from private schools.)

Hoffman's math skills were up to it. "The Paramorphic Representation of Clinical Judgment," he had titled his paper for the *Psychological Bulletin.* If the title was incomprehensible, it was at least in part because Hoffman expected anyone who read it to know what he was talking about. He didn't have any great hope that his paper would be read outside of his small world: What happened in this new little corner of psychology tended to stay there. "People who were making judgments in the real world wouldn't have come across it," said Lew Goldberg. "The people who are not psychologists do not read psychology journals."

The real-world experts whose thinking the Oregon researchers sought to understand were, in the beginning, clinical psychologists, but they clearly believed that whatever they learned would apply more generally to any professional decision maker — doctors, judges, meteorologists, baseball scouts, and so on. "Maybe fifteen people in the world are noodling around on this," said Paul Slovic. "But we recognize we're doing something that could be important: capturing what seemed to be complex, mysterious intuitive judgments with numbers." By the late 1960s Hoffman and his acolytes had reached some unsettling conclusions — nicely captured in a pair of papers written by Lew Goldberg. Goldberg published his first paper in 1968, in an academic journal called *American Psychologist.* He began by pointing out the small mountain of research that suggested that expert judgment was less reliable than algorithms. "I can summarize this ever-growing body of literature," wrote Goldberg, "by pointing out that over a rather large array of clinical judgment tasks (including by now some which were specifically selected to show the clinician at his best and the actuary at his worst), rather simple actuarial formulae typically can be constructed to perform at a level of validity no lower than that of the clinical expert."

So . . . what was the clinical expert doing?

Like others who had approached the problem, Goldberg assumed that when, for instance, a doctor diagnosed a patient, his thinking must be complex. He further assumed that any model seeking to capture that thinking must also be complex. For example, a psychologist at the University of Colorado studying how his fellow psychologists predicted which young people would have trouble adjusting to college had actually taped psychologists talking to themselves as they studied data about their patients — and then tried to write a complicated computer program to mimic the thinking. Goldberg said he preferred to start simple and build from there. As his first case study, he used the way doctors diagnosed cancer.

He explained that the Oregon Research Institute had completed a study of doctors. They had found a gaggle of radiologists at the University of Oregon and asked them: How do you decide from a stomach X-ray if a person has cancer? The doctors said that there were seven major signs that they looked for: the size of the ulcer, the shape of its borders, the width of the crater it made, and so on. The "cues," Goldberg called them, as Hoffman had before him. There were obviously many different plausible combinations of these seven cues, and the doctors had to grapple with how to make sense of them in each of their many combinations. The size of an ulcer might mean one thing if its contours were smooth, for instance, and another if its contours were rough. Goldberg pointed out that, indeed, experts tended to describe their thought processes as subtle and complicated and difficult to model.

The Oregon researchers began by creating, as a starting point, a very simple algorithm, in which the likelihood that an ulcer was malignant depended on the seven factors the doctors had mentioned, equally weighted. The researchers then asked the doctors to judge the probability of cancer in ninety-six different individual stomach ulcers, on a seven-point scale from "definitely malignant" to "definitely benign." Without telling the doctors what they were up to, they showed them each ulcer twice, mixing up the duplicates randomly in the pile so the doctors wouldn't notice they were being asked to diagnose the exact same ulcer they had already diagnosed. The researchers didn't have a computer. They transferred all of their data onto punch cards, which they mailed to UCLA, where the data was analyzed by the university's big computer. The researchers' goal was to see if they could create an algorithm that would mimic the decision making of doctors.

This simple first attempt, Goldberg assumed, was just a starting point. The algorithm would need to become more complex; it would require more advanced mathematics. It would need to account for the subtleties of the doctors' thinking about the cues. For instance, if an ulcer was particularly big, it might lead them to reconsider the meaning of the other six cues.

But then UCLA sent back the analyzed data, and the story became unsettling. (Goldberg described the results as "generally terrifying.") In the first place, the simple model that the researchers had created as their starting point for understanding how doctors rendered their diagnoses proved to be extremely good at predicting the doctors' diagnoses. The doctors might want to believe that their thought processes were subtle and complicated, but a simple model captured these perfectly well. That did not mean that their thinking was necessarily simple, only that it could be captured by a simple model. More surprisingly, the doctors' diagnoses were all over the map: The experts didn't agree with each other. Even more surprisingly, when presented with duplicates of the same ulcer, every doctor had contradicted himself and rendered more than one diagnosis: These doctors apparently could not even agree with themselves. "These findings suggest that diagnostic agreement in clinical medicine may not be much greater than that found in clinical psychology — some food for thought during your next visit to the family doctor," wrote Goldberg. If the doctors disagreed among themselves, they of course couldn't all be right — and they weren't.

The researchers then repeated the experiment with clinical psychologists and psychiatrists, who gave them the list of factors they considered when deciding whether it was safe to release a patient from a psychiatric hospital. Once again, the experts were all over the map. Even more bizarrely, those with the least training (graduate students) were just as accurate as the fully trained ones (paid pros) in their predictions about what any given psychiatric patient would get up to if you let him out the door. Experience appeared to be of little value in judging, say, whether a person was at risk of committing suicide. Or, as Goldberg put it, "Accuracy on this task was not associated with the amount of professional experience of the judge."

Still, Goldberg was slow to blame the doctors. Toward the end of his paper, he suggested that the problem might be that doctors and psychiatrists seldom had a fair chance to judge the accuracy of their thinking and, if necessary, change it. What was lacking was "immediate feedback." And so, with an Oregon Research Institute colleague named Leonard Rorer, he tried to provide it. Goldberg and Rorer gave two groups of psychologists thousands of hypothetical cases to diagnose. One group received immediate feedback on its diagnoses; the other did not — the purpose was to see if the ones who got feedback improved.

The results were not encouraging. "It now appears that our initial formulation of the problem of learning clinical inference was far too simple — that a good deal more than outcome feedback is necessary for judges to learn a task as difficult as this one," wrote Goldberg. At which point one of Goldberg's fellow Oregon researchers — Goldberg doesn't recall which one — made a radical suggestion. "Someone said, 'One of these models you built [to predict what the doctors were doing] might actually be better than the doctor,' " recalled Goldberg. "I thought, Oh, Christ, you idiot, how could that possibly be true?" How could their simple model be better at, say, diagnosing cancer than a doctor? The model had been created, in effect, by the doctors. The doctors had given the researchers all the information in it.

The Oregon researchers went and tested the hypothesis anyway. It turned out to be true. If you wanted to know whether you had cancer or not, you were better off using the algorithm that the researchers had created than you were asking the radiologist to study the X-ray. The simple algorithm had outperformed not merely the group of doctors; it had outperformed even the single best doctor. *You could beat the doctor by replacing him with an equation created by people who knew nothing about medicine and had simply asked a few questions of doctors.*

When Goldberg sat down to write a follow-up paper, which he called "Man versus Model of Man," he was clearly less optimistic than he had formerly been, both about experts and the approach taken by the Oregon Research Institute to understanding their minds. "My article . . . was an account of our experimental failures — failures to demonstrate the complexities of human judgments," he wrote of his earlier piece: the one he'd published in *American*

Psychologist. "Since the previous anecdotal literature was filled with speculations about the complex interactions to be expected when professionals process clinical information, we had naively expected to find that the simple linear combination of cues would not be highly predictive of individual's judgments, and consequently that we would soon be in the business of devising highly complex mathematical expressions to represent individual judgment strategy. Alas, it was not to be." It was as if the doctors had a theory of how much weight to assign to any given trait of any given ulcer. The model captured their theory of how to best diagnose an ulcer. But in practice they did not abide by their own ideas of how to best diagnose an ulcer. As a result, they were beaten by their own model.

The implications were vast. "If these findings can be generalized to other sorts of judgmental problems," Goldberg wrote, "it would appear that only rarely — if at all — will the utilities favor the continued employment of man over a model of man." But how could that be? Why would the judgment of an expert — a medical doctor, no less — be inferior to a model crafted from that very expert's own knowledge? At that point, Goldberg more or less threw up his hands and said, Well, even experts are human. "The clinician is not a machine," he wrote. "While he possesses his full share of human learning and hypothesis-generating skills, he lacks the machine's reliability. He 'has his days': Boredom, fatigue, illness, situational and interpersonal distractions all plague him, with the result that his repeated judgments of the exact same stimulus configuration are not identical. . . . If we could remove some of this human unreliability by eliminating this random error in his judgments, we should thereby increase the validity of the resulting predictions . . ."

Right after Goldberg published those words, late in the summer of 1970, Amos Tversky showed up in Eugene, Oregon. He was on his way to spend a year at Stanford and wanted to visit his old friend Paul Slovic, with whom he'd studied at Michigan. Slovic, a former college basketball player, recalls shooting baskets with Amos in his driveway. Amos, who had not played college basketball, didn't really shoot so much as heave the ball at the rim — his jump shot looked more like calisthenics than hoops. "A three-quarters speed, spinless shot put which started at mid-chest and wafted toward the basket," in the words of his son Oren. And yet Amos had somehow become a basketball enthusiast. "Some people like to walk while they talk. Amos liked to shoot baskets," said Slovic, adding delicately that "he didn't look like someone who had spent a lot of time shooting baskets." Heaving the ball at the rim, Amos told Slovic that he and Danny had been kicking around some ideas about the inner workings of the human mind and hoped to further explore how people made intuitive judgments. "He said they wanted a place where they could just sit and talk to each other all day long without the distraction of a university," said Slovic. They had some thoughts about why even experts might make big, systematic errors. And it wasn't just because they were having a bad day. "And I was just kind of stunned by how exciting the ideas were," said Slovic.

Amos had agreed to spend the 1970–71 academic year at Stanford University, and so he and Danny, who remained in Israel, were apart. They used the year to collect data. The data consisted entirely of answers to curious questions that they had devised. Their questions were first posed to high school students in Israel — Danny sent out twenty or so Hebrew University graduate students in taxis to scour the entire country for unsuspecting Israeli children. ("We were running out of kids in Jerusalem.") The graduate students gave each kid two to four of what must have seemed to them totally bizarre questions, and a couple of minutes to answer each of them. "We had multiple questionnaires," said Danny, "because no one child could do the whole thing."

> Consider the following question:
> All families of six children in a city were surveyed. In 72 families the exact order of births of boys and girls was G B G B B G.
>
> What is your estimate of the number of families surveyed in which the exact order of births was B G B B B B?

That is, in this hypothetical city, if there were 72 families with 6 children born in the following order — girl, boy, girl, boy, boy, girl — how many families with 6 children do you imagine have the birth order boy, girl, boy, boy, boy, boy? Who knows what Israeli high school students made of the strange question, but fifteen hundred of them supplied answers to it. Amos posed other, equally weird, questions to college students at the University of Michigan and Stanford University. For example:

On each round of a game, 20 marbles are distributed at random among five children: Alan, Ben, Carl, Dan, and Ed. Consider the following distributions:

I	II
Alan, 4	Alan, 4
Ben, 4	Ben, 4
Carl, 5	Carl, 4
Dan, 4	Dan, 4
Ed, 3	Ed, 4

In many rounds of the game, will there be more results of type I or type II?

They were trying to determine how people judged — or, rather, misjudged — the odds of any situation when the odds were hard, or impossible, to know. All the questions had right answers and wrong answers. The answers that their subjects supplied could be compared to the right answer, and their errors investigated for patterns. "The general idea was: What do people do?" said Danny. "What actually is going on when people judge probability? It's a very abstract concept. They must be doing something."

Amos and Danny didn't have much doubt that a lot of people would get the questions they had dreamed up wrong — because Danny and Amos had gotten them, or versions of them, wrong. More precisely, Danny made the mistakes, noticed that he made the mistakes, and theorized about why he had made the mistakes, and Amos became so engrossed by both Danny's mistakes and his perceptions of those mistakes that he at least pretended to have been tempted to make the same ones. "We kicked it around, and our focus became *our* intuitions," said Danny. "We thought that errors we did not make ourselves were not interesting." If they both committed the same mental errors, or were tempted to commit them, they assumed — rightly, as it turned out — that most other people would commit them, too. The questions they had spent the year cooking up for the students in Israel and the United States were not so much experiments as they were little dramas: *Here, look, this is what the uncertain human mind actually does.*

At a very young age, Amos had recognized a distinction within the class of people who insisted on making their lives complicated. Amos had a gift for avoiding what he called "overcomplicated" people. But every now and then he ran into a person, usually a woman, whose complications genuinely interested him. In high school he'd become entranced with the future poet Dahlia Ravikovitch: His intimate friendship with her had startled their peers. His relationship with Danny had the same effect. An old friend of Amos's would later recall, "Amos would say, 'People are not so complicated. *Relationships* between people are complicated.' And then he would pause, and say: 'Except for Danny.' " But there was something about Danny that caused Amos to let down his guard and turned Amos, when he was alone with Danny, into a different character. "Amos almost suspended disbelief when we were working together," said Danny. "He didn't do that much for other people. And *that* was the engine of the collaboration."

In August 1971 Amos returned to Eugene with his wife and children and a mental pile of data, and moved into a house on a cliff overlooking the town. He'd rented it from an Oregon Research Institute psychologist on leave. "The thermostat was set on 85," said Barbara. "There were picture windows, with no curtain. They had left a mountain of laundry, none of it clothes." Their landlords, they soon learned, were nudists. (Welcome to Eugene! Don't look down!) A few weeks later Danny followed with his own wife and children, and an even bigger mental pile of data, and moved into a house with something even more unsettling — to Danny — than a nudist: a lawn. Danny couldn't picture himself doing yard work any more than anyone else could picture him doing it. Still, he was unusually optimistic. "My memories of Eugene are all of bright sunshine," he later said, even though he had come from a land where the sun shined all the time, and, on more than half the days he spent in Eugene, the skies were more cloudy than blue.

Anyway, he spent most of his time indoors, talking to Amos. They installed themselves in an office in the former Unitarian church, and continued the conversation they'd started in Jerusalem. "I had the sense, 'My life has changed,' " said Danny. "We were quicker in understanding each other than we were in understanding ourselves. The way the creative process works is that you first say something, and later, sometimes years later, you understand what you said. And in our case it was foreshortened. I would say something and Amos would understand it. When one of us would say something that was off the wall, the other would search for the virtue in it. We would finish each other's sentences and frequently did. But we also kept surprising each other. It still gives me goose bumps." For the first time in their careers, they had something like a staff at their disposal. Papers got typed by someone else; subjects for their experiments got recruited by someone else; money for research got raised by someone else. All they had to do was talk to each other.

They had some ideas about the mechanisms in the human mind that produced error. They set out looking for the interesting mistakes — or biases — that such mechanisms would make. A pattern emerged: Danny would arrive early each morning and analyze the answers that Oregon college students had given to their questions of the day before. (Danny didn't believe in waiting around: He'd later admonish graduate students who failed to analyze data within a day of getting it, saying, "It's a bad sign for your research career.") Amos would turn up around noon and the two of them would walk down to a fish and chips place no one else could stand, eat lunch, and then return and talk the rest of the day. "They had a certain style of working," recalls Paul Slovic, "which is they just talked to each other for hour after hour after hour."

The Oregon researchers noticed, as the Hebrew University professors had noticed, that whatever Amos and Danny were talking about must be funny, as they spent half their time laughing. They bounced back and forth between Hebrew and English and broke each other up in both. They happened to be in Eugene, Oregon, surrounded by joggers and nudists and hippies and forests of Ponderosa pine, but they could just as well have been in Mongolia. "I don't think either of them was attached to physical location," said Slovic. "It didn't matter where they were. All that mattered were the ideas." Everyone also noticed the intense privacy of their conversation. Before they had arrived in Eugene, Amos had made some faint noises about including Paul Slovic in the collaboration, but once Danny arrived it became clear to Slovic that he didn't belong. "We weren't a threesome together much," he said. "They didn't want anyone else in the room."

In a funny way, they didn't even want themselves in the room. They wanted to be the people they became when they were with each other. Work, for Amos, had always been play: If it wasn't fun, he simply didn't see the point in doing it. Work now became play for Danny, too. This was new. Danny was like a kid with the world's best toy closet who is so paralyzed by indecision that he never gets around to enjoying his possessions but instead just stands there worrying himself to death over whether to grab his Super Soaker or take his electric scooter out for a spin. Amos rooted around in Danny's mind and said, "Screw it, we're going to play with *all of this stuff.*" There would be times, later in their relationship, when Danny would go into a deep funk — a depression, almost — and walk around saying, "I'm out of ideas." Even that Amos found funny. Their mutual friend Avishai Margalit recalled, "When he heard that Danny was saying, 'I'm finished, I'm out of ideas,' Amos laughed and said, 'Danny has more ideas in one minute than a hundred people have in a hundred years.' " When they sat down to write they nearly merged, physically, into a single form, in a way that the few people who happened to catch a glimpse of them found odd. "They wrote together sitting right next to each other at the typewriter," recalls Michigan psychologist Richard Nisbett. "I cannot imagine. It would be like having someone else brush my teeth for me." The way Danny put it was, "We were sharing a mind."

Their first paper — which they still half-thought of as a joke played on the academic world — had shown that people faced with a problem that had a statistically correct answer did not think like statisticians. Even statisticians did not think like statisticians. "Belief in the Law of Small Numbers" had raised an obvious next question: If people did not use statistical reasoning, even when faced with a problem that could be solved with statistical reasoning, what kind of reasoning did they use? If they did not think, in life's many chancy situations, like a card counter at a blackjack table, how did they think? Their next paper offered a partial answer to the question. It was called . . . well, Amos had this thing about titles. He refused to start a paper until he had decided what it would be called. He believed the title forced you to come to grips with what your paper was about.

And yet the titles that he and Danny put on their papers were inscrutable. They had to play, at least in the beginning, by the rules of the academic game, and in that game it wasn't quite respectable to be easily understood. Their first attempt to describe how people formed judgments they titled "Subjective Probability: A Judgment of Representativeness."* *Subjective probability*

* Having realized at the start of their collaboration that they would never be able to work out who had contributed more to any given paper, they alternated lead authorship. Because Amos

— a person might just make out what that meant. Subjective probability meant: the odds you assign to any given situation when you are more or less guessing. Look outside the window at midnight and see your teenage son weaving his way toward your front door, and say to yourself, "There's a 75 percent chance he's been drinking" — that's subjective probability. But "A Judgment of Representativeness": What the hell was that? "Subjective probabilities play an important role in our lives," they began. "The decisions we make, the conclusions we reach, and the explanations we offer are usually based on our judgments of the likelihood of uncertain events such as success in a new job, the outcome of an election, or the state of a market." In these and many other uncertain situations, the mind did not naturally calculate the correct odds. So what did it do?

The answer they now offered: It replaced the laws of chance with rules of thumb. These rules of thumb Danny and Amos called "heuristics." And the first heuristic they wanted to explore they called "repre-

---

had won the coin flip to be lead author on "Belief in the Law of Small Numbers," Danny was lead author on this new paper.

sentativeness."

When people make judgments, they argued, they compare whatever they are judging to some model in their minds. How much do those clouds resemble my mental model of an approaching storm? How closely does this ulcer resemble my mental model of a malignant cancer? Does Jeremy Lin match my mental picture of a future NBA player? Does that belligerent German political leader resemble my idea of a man capable of orchestrating genocide? The world's not just a stage. It's a casino, and our lives are games of chance. And when people calculate the odds in any life situation, they are often making judgments about similarity — or (strange new word!) representativeness. You have some notion of a parent population: "storm clouds" or "gastric ulcers" or "genocidal dictators" or "NBA players." You compare the specific case to the parent population.

Amos and Danny left unaddressed the question of how exactly people formed mental models in the first place, and how they made judgments of similarity. Instead, they said, let's focus on cases where the mental model that people have in their heads is fairly obvious. The more similar the specific case is to the notion in your head, the more likely you are to believe that the case belongs to the larger group. "Our thesis," they wrote, "is that, in many situations, an event *A* is judged to be more probable than an event *B* whenever *A* appears more representative than *B*." The more the basketball player resembles your mental model of an NBA player, the more likely you will think him to be an NBA player.

They had a hunch that people, when they formed judgments, weren't just making random mistakes — that they were doing something systematically wrong. The weird questions they put to Israeli and American students were designed to tease out the pattern in human error. The problem was subtle. The rule of thumb they had called representativeness wasn't always wrong. If the mind's approach to uncertainty was occasionally misleading, it was because it was often so useful. Much of the time, the person who can become a good NBA player matches up pretty well with the mental model of "good NBA player." But sometimes a person does not — and in the systematic errors they led people to make, you could glimpse the nature of these rules of thumb.

For instance, in families with six children, the birth order B G B B B B was about as likely as G B G B B G. But Israeli kids — like pretty much everyone else on the planet, it would emerge — naturally seemed to believe that G B G B B G was a more likely birth sequence. Why? "The sequence with five boys and one girl fails to reflect the proportion of boys and girls in the population," they explained. It was less representative. What is more, if you asked the same Israeli kids to choose the more likely birth order in families with six children — B B B G G G or G B B G B G — they overwhelmingly opted for the latter. But the two birth orders are equally likely. So why did people almost universally believe that one was far more likely than the other? Because, said Danny and Amos, people thought of birth order as a random process, and the second sequence looks more "random" than the first.

The natural next question: When does our rule-of-thumb approach to calculating the odds lead to serious miscalculation? One answer was: Whenever people are asked to evaluate anything with a random component to it. It wasn't enough that the uncertain event being judged resembled the parent population, wrote Danny and Amos. "The event should also reflect the properties of the uncertain process by which it is generated." That is, if a process is random, its outcome should appear random. They didn't explain how people's mental model of "randomness" was formed in the first place. Instead they said, *Let's look at judgments that involve randomness, because we psychologists can all pretty much agree on people's mental model of it.*

Londoners in the Second World War thought that German bombs were targeted, because some parts of the city were hit repeatedly while others were not hit at all. (Statisticians later showed that the distribution was exactly what you would expect from random bombing.) People find it a remarkable coincidence when two students in the same classroom share a birthday, when in fact there is a better than even chance, in any group of twenty-three people, that two of its members will have been born on the same day. We have a kind of stereotype of "randomness" that differs from true randomness. Our stereotype of randomness lacks the clusters and patterns that occur in true random sequences. If you pass out twenty marbles randomly to five boys, they are actually more likely to each receive four marbles (column II), than they are to receive the combination in column I, and yet American college students insisted that the unequal distribution in column I was more likely than the equal one in column II. Why? Because column II "appears too lawful to be the result of a random process. . . ."

A suggestion arose from Danny and Amos's paper: If our minds can be misled by our false stereotype of something as measurable as randomness, how much might they be misled by other, vaguer stereotypes?

> The average heights of adult males and females in the U.S. are, respectively, 5 ft. 10 in. and 5 ft. 4 in. Both distributions are approximately normal with a standard deviation of about 2.5 in.*
>
> An investigator has selected one population by chance and has drawn from it a

* Standard deviation is a measurement of the dispersal of any population. The bigger the standard deviation, the more varied the population. A standard deviation of 2.5 inches in a world in which the average man is five foot ten means that roughly 68 percent of men are between 5 feet 7-1/2 inches and six feet 1/2 inch. If the standard deviation was zero, all men would be exactly five foot ten.

random sample.

What do you think the odds are that he has selected the male population if

1. The sample consists of a single person whose height is 5 ft. 10 in.?
2. The sample consists of 6 persons whose average height is 5 ft. 8 in.?

The odds most commonly assigned by their subjects were, in the first case, 8:1 in favor and, in the second case, 2.5:1 in favor. The correct odds were 16:1 in favor in the first case, and 29:1 in favor in the second case. The sample of six people gave you a lot more information than the sample of one person. And yet people believed, incorrectly, that if they picked a single person who was five foot ten, they were more likely to have picked from the population of men than had they picked six people with an average height of five foot eight. People didn't just miscalculate the true odds of a situation: They treated the less likely proposition as if it were the more likely one. And they did this, Amos and Danny surmised, because they saw "5 ft. 10 in." and thought: That's the typical guy! The stereotype of the man blinded them to the likelihood that they were in the presence of a tall woman.

> A certain town is served by two hospitals. In the larger hospital about 45 babies are born each day, and in the smaller hospital about 15 babies are born each day. As you know, about 50 percent of all babies are boys. The exact percentage of baby boys, however, varies from day to day. Sometimes it may be higher than 50 percent, sometimes lower.
>
> For a period of 1 year, each hospital recorded the days on which more than 60 percent of the babies born were boys. Which hospital do you think recorded more such days? Check one:
>
> ________ The larger hospital
> ________ The smaller hospital
> ________ About the same (that is, within 5 percent of each other)

People got that one wrong, too. Their typical answer was "same." The correct answer is "the smaller hospital." The smaller the sample size, the more likely that it is unrepresentative of the wider population. "We surely do not mean to imply that man is incapable of appreciating the impact of sample size on sampling variance," wrote Danny and Amos. "People can be taught the correct rule, perhaps even with little difficulty. The point remains that people do not follow the correct rule, when left to their own devices."

To which a bewildered American college student might reply: All these strange questions! What do they have to do with my life? A great deal, Danny and Amos clearly believed. "In their daily lives," they wrote, "people ask themselves and others questions such as: What are the chances that this 12-year-old boy will grow up to be a scientist? What is the probability that this candidate will be elected to office? What is the likelihood that this company will go out of business?" They confessed that they had confined their questions to situations in which the odds could be objectively calculated. But they felt fairly certain that people made the same mistakes when the odds were harder, or even impossible, to know. When, say, they guessed what a little boy would do for a living when he grew up, they thought in stereoypyes. If he matched their mental picture of a scientist, they guessed he'd be a scientist — and neglect the prior odds of any kid becoming a scientist.

Of course, you couldn't prove that people

misjudged the odds of a situation when the odds were extremely difficult or even impossible to know. How could you prove that people came to the wrong answer when a right answer didn't exist? But if people's judgments were distorted by representativeness when the odds were knowable, how likely was it that their judgments were any better when the odds were a total mystery?

Danny and Amos had their first big general idea — the mind had these mechanisms for making judgments and decisions that were usually useful but also capable of generating serious error. The next paper they produced inside the Oregon Research Institute described a second mechanism, an idea that had come to them just a couple of weeks after the first. "It wasn't all representativeness," said Danny. "There was something else going on. It wasn't just similarity." The new paper's title was once again more mystifying than helpful: "Availability: A Heuristic for Judging Frequency and Probability." Once again, the authors came with news of the results of questions that they had posed to students, mostly at the University of Oregon, where they now had an endless supply of lab rats. They'd gathered a lot more kids in classrooms and asked them, absent a dictionary or any text, to answer these bizarre questions:

> The frequency of appearance of letters in the English language was studied. A typical text was selected, and the relative frequency with which various letters of the alphabet appeared in the first and third positions of the words was recorded. Words of less than three letters were excluded from the count.
>
> You will be given several letters of the alphabet, and you will be asked to judge whether these letters appear more often in the first or in the third position, and to estimate the ratio of the frequency with which they appear in these positions. . . .
>
> Consider the letter *K*
>
> Is *K* more likely to appear in
>
> ________ the first position?
> ________ the third position? (check one)
>
> My estimate for the ratio of these two values is: ________:1

If you thought that *K* was, say, twice as likely to appear as the first letter of an

English word than as the third letter, you checked the first box and wrote your estimate as 2:1. This was what the typical person did, as it happens. Danny and Amos replicated the demonstration with other letters — *R, L, N,* and *V.* Those letters all appeared more frequently as the third letter in an English word than as the first letter — by a ratio of two to one. Once again, people's judgment was, systematically, very wrong. And it was wrong, Danny and Amos now proposed, because it was distorted by memory. It was simply easier to recall words that start with *K* than to recall words with *K* as their third letter.

The more easily people can call some scenario to mind — the more *available* it is to them — the more probable they find it to be. Any fact or incident that was especially vivid, or recent, or common — or anything that happened to preoccupy a person — was likely to be recalled with special ease, and so be disproportionately weighted in any judgment. Danny and Amos had noticed how oddly, and often unreliably, their own minds recalculated the odds, in light of some recent or memorable experience. For instance, after they drove past a gruesome car crash on the highway, they slowed down: Their sense of the odds of being in a crash had changed. After seeing a movie that dramatizes nuclear war, they worried more about nuclear war; indeed, they felt that it was more likely to happen. The sheer volatility of people's judgment of the odds — their sense of the odds could be changed by two hours in a movie theater — told you something about the reliability of the mechanism that judged those odds.

They went on to describe nine other equally odd mini-experiments that got at various tricks that memory might play on judgment. Danny thought of them as very much like the optical illusions the Gestalt psychologists he had loved in his youth planted in their texts. You saw them and were fooled by them and wanted to know why. He and Amos were dramatizing tricks of the mind rather than tricks of the eye, but the effect was similar, and the material available to them appeared to be even more abundant. They read lists of people's names to Oregon students, for instance. Thirty-nine names, read at a rate of two seconds per name. The names were all easily identifiable as male or female. A few were the names of famous people — Elisabeth Taylor, Richard Nixon. A few were names of slightly less famous people — Lana Turner,

William Fulbright. One list consisted of nineteen male names and twenty female names, the other of twenty female names and nineteen male names. The list that had more female names on it had more names of famous men, and the list that had more male names on it contained the names of more famous women. The unsuspecting Oregon students, having listened to a list, were then asked to judge if it contained the names of more men or more women.

They almost always got it backward: If the list had more male names on it, but the women's names were famous, they thought the list contained more female names, and vice versa. "Each of the problems had an objectively correct answer," Amos and Danny wrote, after they were done with their strange mini-experiments. "This is not the case in many real-life situations where probabilities are judged. Each occurrence of an economic recession, a successful medical operation, or a divorce, is essentially unique, and its probability cannot be evaluated by a simple tally of instances. Nevertheless, the availability heuristic may be applied to evaluate the likelihood of such events. "In judging the likelihood that a particular couple will be divorced, for example, one may scan one's memory for similar couples which this question brings to mind. Divorces will appear probable if divorces are prevalent among the instances that are retrieved in this manner."

The point, once again, wasn't that people were stupid. This particular rule they used to judge probabilities (the easier it is for me to retrieve from my memory, the more likely it is) often worked well. But if you presented people with situations in which the evidence they needed to judge them accurately was hard for them to retrieve from their memories, and misleading evidence came easily to mind, they made mistakes. "Consequently," Amos and Danny wrote, "the use of the availability heuristic leads to systematic biases." Human judgment was distorted by . . . the *memorable.*

Having identified what they took to be two of the mind's mechanisms for coping with uncertainty, they naturally asked: Are there others? Apparently they were unsure. Before they left Eugene, they jotted down some notes about other possibilities. "The conditionality heuristic," they called one of these. In judging the degree of uncertainty in any situation, they noted, people made "unstated assumptions." "In assessing the profit of a given company, for example, people tend to assume normal operating conditions and make their estimates contingent upon that assumption," they wrote in their notes. "They do not incorporate into their estimates the possibility that these conditions may be drastically changed because of a war, sabotage, depressions, or a major competitor being forced out of business." Here, clearly, was another source of error: not just that people don't know what they don't know, but that they don't bother to factor their ignorance into their judgments.

Another possible heuristic they called "anchoring and adjustment." They first dramatized its effects by giving a bunch of high school students five seconds to guess the answer to a math question. The first group was asked to estimate this product:

$$8 \times 7 \times 6 \times 5 \times 4 \times 3 \times 2 \times 1$$

The second group to estimate this product:

$$1 \times 2 \times 3 \times 4 \times 5 \times 6 \times 7 \times 8$$

Five seconds wasn't long enough to actually do the math: The kids had to guess. The two groups' answers should have been at least roughly the same, but they weren't, even roughly. The first group's median answer was 2,250. The second group's median answer was 512. (The right answer is 40,320.) The reason the kids in the first group guessed a higher number for the first sequence was that they had used 8 as a starting point, while the kids in the second group had used 1.

It was almost too easy to dramatize this weird trick of the mind. People could be anchored with information that was totally irrelevant to the problem they were being asked to solve. For instance, Danny and Amos asked their subjects to spin a wheel of fortune with slots on it that were numbered 0 through 100. Then they asked the subjects to estimate the percentage of African countries in the United Nations. The people who spun a higher number on the wheel tended to guess that a higher percentage of the United Nations consisted of African countries than did those for whom the needle landed on a lower number. What was going on here? Was anchoring a heuristic, the way that representativeness and availability were heuristics? Was it a shortcut that people used, in effect, to answer to their own satisfaction a question to which they could not divine the true answer? Amos thought it was; Danny thought it wasn't. They never came to sufficient agreement to write a paper on the subject. Instead they dropped it into summaries of their work. "We had to stick anchoring in, because the result was so spectacular," said Danny. "But as a result we wound up with a vague notion of what a heuristic is."

Danny would later say that it was hard to explain what he and Amos were doing in the beginning: "How can you explain a conceptual fog?" he said. "We didn't have the intellectual tools to understand what we were finding." Were they investigating the biases or the heuristics? The errors, or the mechanisms that produced the errors? The errors enabled you to offer at least a partial description of the mechanism: The bias was the footprint of the heuristic. The biases, too, would soon have their own names, like the "recency bias" and the "vividness bias." But in hunting for errors that they themselves had made, and then tracking them back to their source in the human mind, they had stumbled upon errors without a visible trail. What were they to make of systematic errors for which there was no apparent mechanism? "We really couldn't think of others," said Danny. "There seemed to be very few mechanisms."

Just as they never tried to explain how the mind forms the models that underpinned the representativeness heuristic, they left mostly to one side the question of why human memory worked in such a way that the availability heuristic had such power to mislead us. They focused entirely on the various tricks it could play. The more complicated and lifelike the situation a person was asked to judge, they suggested, the more insidious the role of availability. What people did in many complicated real-life problems — when trying to decide if Egypt might invade Israel, say, or their husband might leave them for another woman — was to construct scenarios. The stories we make up, rooted in our memories, effectively replace probability judgments. "The production of a compelling scenario is likely to constrain future thinking," wrote Danny and Amos. "There is much evidence showing that, once an uncertain situation has been perceived or interpreted in a particular fashion, it is quite difficult to view it in any other way."

But these stories people told themselves were biased by the availability of the material used to construct them. "Images of the future are shaped by experience of the past," they wrote, turning on its head Santayana's famous lines about the importance of history: *Those who cannot remember the past are condemned to repeat it.* What people remember about the past, they suggested, is likely to warp their judgment of the future. "We often decide that an outcome is extremely unlikely or impossible, because we are unable to imagine any chain of events that could cause it to occur. The defect, often, is in our imagination."*

The stories people told themselves, when the odds were either unknown or unknowable, were naturally too simple. "This tendency to consider only relatively simple scenarios," they concluded, "may have particularly salient effects in situations of conflict. There, one's own moods and plans are more available to one than those of the opponent. It is not easy to adopt the opponent's view of the chessboard or of the battlefield." The imagination appeared to be governed by rules. The rules confined people's thinking. It's far easier for a Jew living in Paris in 1939 to construct a story about how the German army will behave much as it had in 1919, for instance, than to invent a story in which it behaves as it did in 1941, no matter how persuasive the

* Those lines come not from their published paper but from a summary of their work that they produced a year after the paper's publication.

evidence might be that, this time, things are different.

7
The Rules of Prediction

Amos liked to say that if you are asked to do anything — go to a party, give a speech, lift a finger — you should never answer right away, even if you are sure that you want to do it. *Wait a day,* Amos said, *and you'll be amazed how many of those invitations you would have accepted yesterday you'll refuse after you have had a day to think it over.* A corollary to his rule for dealing with demands upon his time was his approach to situations from which he wished to extract himself. A human being who finds himself stuck at some boring meeting or cocktail party often finds it difficult to invent an excuse to flee. Amos's rule, whenever he wanted to leave any gathering, was to just get up and leave. *Just start walking and you'll be surprised how creative you will become and how fast you'll find the words for your excuse,* he said. His attitude to the clutter of daily life was of a piece with his strategy for dealing with social demands. *Unless you are kicking yourself once a month for throwing something away, you are not throwing enough away,* he said. Everything that didn't seem to Amos obviously important he chucked, and thus what he saved acquired the interest of objects that have survived a pitiless culling. One unlikely survivor is a single scrap of paper with a few badly typed words on it, drawn from conversations he had with Danny in the spring of 1972 as they neared the end of their time in Eugene. For some reason Amos saved it:

People predict by making up stories
People predict very little and explain everything
People live under uncertainty whether they like it or not
People believe they can tell the future if they work hard enough
People accept any explanation as long as it fits the facts
The handwriting was on the wall, it was just the ink that was invisible

People often work hard to obtain information they already have And avoid new knowledge
Man is a deterministic device thrown into

a probabilistic Universe
In this match, surprises are expected
Everything that has already happened
must have been inevitable

At first glance it resembles a poem. What it was, in fact, was early fodder for his and Danny's next article, which would also be their first attempt to put their thinking in such a way that it might directly influence the world outside of their discipline. Before returning to Israel, they had decided to write a paper about how people made predictions. The difference between a judgment and a prediction wasn't as obvious to everyone as it was to Amos and Danny. To their way of thinking, a judgment ("he looks like a good Israeli army officer") implies a prediction ("he will make a good Israeli army officer"), just as a prediction implies some judgment — without a judgment, how would you predict? In their minds, there was a distinction: A prediction is a judgment that involves uncertainty. "Adolf Hitler is an eloquent speaker" is a judgment you can't do much about. "Adolf Hitler will become chancellor of Germany" is, at least until January 30, 1933, a prediction of an uncertain event that eventually will be proven either right or wrong. The title of their next paper was "On the Psychology of Prediction." "In making predictions and judgments under uncertainty," they wrote, "people do not appear to follow the calculus of chance or the statistical theory of prediction. Instead, they rely on a limited number of heuristics which sometimes yield reasonable judgments and sometimes lead to severe and systematic error."

Viewed in hindsight, the paper looks to have more or less started with Danny's experience in the Israeli army. The people in charge of vetting Israeli youth hadn't been able to predict which of them would make good officers, and the people in charge of officer training school hadn't been able to predict who among the group they were sent would succeed in combat, or even in the routine day-to-day business of leading troops. Danny and Amos had once had a fun evening trying to predict the future occupations of their friends' small children, and had surprised themselves by the ease, and the confidence, with which they had done it. Now they sought to test how people predicted — or, rather, to dramatize how people used what they now called the representativeness heuristic to predict.

To do this, however, they needed to give them something to predict.

They decided to ask their subjects to predict the future of a student, identified only by some personality traits, who would go on to graduate school. Of the then nine major courses of graduate study in the United States, which would he pursue? They began by asking their subjects to estimate the percentage of students in each course of study. Here were their average guesses:

Business: 15 percent Computer
Science: 7 percent
Engineering: 9 percent
Humanities and Education: 20 percent
Law: 9 percent
Library Science: 3 percent
Medicine: 8 percent
Physical and Life Sciences: 12 percent
Social Science and Social Work: 17 percent

For anyone trying to predict which area of study any given person was in, those percentages should serve as a base rate. That is, if you knew nothing at all about a particular student, but knew that 15 percent of all graduate students were pursuing degrees in business administration, and were asked to predict the likelihood that the student in question was in business school, you should guess "15 percent." Here was a useful way of thinking about base rates: They were what you would predict if you had no information at all.

Now Danny and Amos sought to dramatize what happened when you gave people some information. But what kind of information? Danny spent a day inside the Oregon Research Institute stewing over the question — and became so engrossed by his task that he stayed up all night creating what at the time seemed like the stereotype of a graduate student in computer science. He named him "Tom W."

> Tom W. is of high intelligence, although lacking in true creativity. He has a need for order and clarity, and for neat and tidy systems in which every detail finds its appropriate place. His writing is rather dull and mechanical, occasionally enlivened by somewhat corny puns and by flashes of imagination of the sci-fi type. He has a strong drive for competence. He seems to have little feel and little sympathy for other people and does not enjoy interacting with others. Self-centered, he nonetheless has a deep moral sense.

They would ask one group of subjects —

they called it the "similarity" group — to estimate how "similar" Tom was to the graduate students in each of the nine fields. That was simply to determine which field of study was most "representative" of Tom W.

Then they would hand a second group — what they called the "prediction" group — this additional information:

> The preceding personality sketch of Tom W. was written during Tom's senior year in high school by a psychologist, on the basis of projective tests. Tom W. is currently a graduate student. Please rank the following nine fields of graduate specialization in order of the likelihood that Tom W. is now a graduate student in each of these fields.

They would not only give their subjects the sketch but inform them that it was a far from reliable description of Tom W. That it had been written by a psychologist, for a start; they would further tell subjects that the assessment had been made years earlier. What Amos and Danny suspected — because they had tested it first on themselves — is that people would essentially leap from the similarity judgment ("that guy sounds like a computer scientist!") to some prediction ("that guy must be a computer scientist!") and ignore both the base rate (only 7 percent of all graduate students were computer scientists) and the dubious reliability of the character sketch.

The first person to arrive for work on the morning Danny finished his sketch was an Oregon researcher named Robyn Dawes. Dawes was trained in statistics and legendary for the rigor of his mind. Danny handed him the sketch of Tom W. "He read it over and he had a sly smile, as if he had figured it out," said Danny. "And he said, 'Computer scientist!' After that I wasn't worried about how the Oregon students would fare."

The Oregon students presented with the problem simply ignored all objective data and went with their gut sense, and predicted with great certainty that Tom W. was a computer scientist. Having established that people would allow a stereotype to warp their judgment, Amos and Danny then wondered: If people are willing to make irrational predictions based on that sort of information, what kind of predictions might they make if we give them totally irrelevant information? As they played with this idea — they might increase people's confidence in their predictions by giving them any information, however useless — the laughter to be heard from the other side of the closed door must have grown only more raucous. In the end, Danny created another character. This one he named "Dick":

> Dick is a 30 year old man. He is married with no children. A man of high ability and high motivation, he promises to be quite successful in his field. He is well liked by his colleagues.

Then they ran another experiment. It was a version of the book bag and poker chips experiment that Amos and Danny had argued about in Danny's seminar at Hebrew University. They told their subjects that they had picked a person from a pool of 100 people, 70 of whom were engineers and 30 of whom were lawyers. Then they asked them: What is the likelihood that the selected person is a lawyer? The subjects correctly judged it to be 30 percent. And if you told them that you were doing the same thing, but from a pool that had 70 lawyers in it and 30 engineers, they said, correctly, that there was a 70 percent chance the person you'd plucked from it was a lawyer. But if you told them you had picked not just some nameless person but a guy named

Dick, and read them Danny's description of Dick — *which contained no information whatsoever to help you guess what Dick did for a living* — they guessed there was an equal chance that Dick was a lawyer or an engineer, no matter which pool he had emerged from. "Evidently, people respond differently when given no specific evidence and when given worthless evidence," wrote Danny and Amos. "When no specific evidence is given, the prior probabilities are properly utilized; when worthless specific evidence is given, prior probabilities are ignored."*

There was much more to "On the Psychology of Prediction" — for instance, they showed that the very factors that caused people to become more confident in their

* By the time they were finished with the project, they had dreamed up an array of hysterically bland characters for people to evaluate and judge to be more likely lawyers or engineers. Paul, for example. "Paul is 36 years old, married, with 2 children. He is relaxed and comfortable with himself and with others. An excellent member of a team, he is constructive and not opinionated. He enjoys all aspects of his work, and in particular, the satisfaction of finding clean solutions to complex problems."

predictions also led those predictions to be less accurate. And in the end it returned to the problem that had interested Danny since he had first signed on to help the Israeli army rethink how it selected and trained incoming recruits:

> The instructors in a flight school adopted a policy of consistent positive reinforcement recommended by psychologists. They verbally reinforced each successful execution of a flight maneuver. After some experience with this training approach, the instructors claimed that contrary to psychological doctrine, high praise for good execution of complex maneuvers typically results in a decrement of performance on the next try. What should the psychologist say in response?

The subjects to whom they posed this question offered all sorts of advice. They surmised that the instructors' praise didn't work because it led the pilots to become overconfident. They suggested that the instructors didn't know what they were talking about. No one saw what Danny saw: that the pilots would have tended to do better after an especially poor maneuver, or worse after an especially great one, if no one had said anything at all. Man's inability to see the power of regression to the mean leaves him blind to the nature of the world around him. *We are exposed to a lifetime schedule in which we are most often rewarded for punishing others, and punished for rewarding.*

When they wrote their first papers, Danny and Amos had no particular audience in mind. Their readers would be the handful of academics who happened to subscribe to the highly specialized psychology trade journals in which they published. By the summer of 1972, they had spent the better part of three years uncovering the ways in which people judged and predicted — but the examples that they had used to illustrate their ideas were all drawn directly from psychology, or from the strange, artificial-seeming tests that they had given high school and college students. Yet they were certain that their insights applied anywhere in the world that people were judging probabilities and making decisions. They sensed that they needed to find a broader audience. "The next phase of the project will be devoted primarily to the extension and application of this work to other high-level professional activities, e.g., economic planning, technological forecasting, political decision making, medical diagnosis, and the evaluation of legal evidence," they wrote in a research proposal. They hoped, they wrote, that the decisions made by experts in these fields could be "significantly improved by making these experts aware of their own biases, and by the development of methods to reduce and counteract the sources of bias in judgment." They wanted to turn the real world into a laboratory. It was no longer just students who would be their lab rats but also doctors and judges and politicians. The question was: How to do it?

They couldn't help but sense, during their year in Eugene, a growing interest in their work. "That was the year it was really clear we were onto something," recalled Danny. "People started treating us with respect." Irv Biederman, then a visiting associate professor of psychology at Stanford University, heard Danny give a talk about heuristics and biases on the Stanford campus in early 1972. "I remember I came home from the talk and told my wife, 'This is going to win a Nobel Prize in economics,' " recalled Biederman. "I was so absolutely convinced. This was a psychological theory about economic man. I thought, What could be better? *Here* is why you get all these irrationalities and errors. They come from the inner workings of the human mind."

Biederman had been friends with Amos at the University of Michigan and was now a member of the faculty at the State University of New York at Buffalo. The Amos he knew was consumed by possibly important but probably insolvable and certainly obscure problems about measurement. "I wouldn't have invited Amos to Buffalo to talk about that," he said — as no one would have understood it or cared about it. But this new work Amos was apparently doing with Danny Kahneman was breathtaking. It confirmed Biederman's sense that "most advances in science come not from eureka moments but from *'hmmm, that's funny.'* " He persuaded Amos to pass through Buffalo in the summer of 1972, on his way from Oregon to Israel. Over the course of a week, Amos gave five different talks about his work with Danny, each aimed at a different group of academics. Each time, the room was jammed — and fifteen years later, in 1987, when Biederman left Buffalo for the University of Minnesota, people were still talking about Amos's talks.

Amos devoted talks to each of the heuristics he and Danny had discovered, and another to prediction. But the talk that lingered in Biederman's mind was the fifth and final one. "Historical Interpretation: Judgment Under Uncertainty," Amos had called it. With a flick of the wrist, he showed a roomful of professional historians just how much of human experience could be reexamined in a fresh, new way, if seen through the lens he had created with Danny.

> In the course of our personal and professional lives, we often run into situations that appear puzzling at first blush. We cannot see for the life of us why Mr. X acted in a particular way, we cannot understand how the experimental results came out the way they did, etc. Typically, however, within a very short time we come up with an explanation, a hypothesis, or an interpretation of the facts that renders them understandable, coherent, or natural. The same phenomenon is observed in perception. People are very good at detecting patterns and trends even in random data. In contrast to our skill in inventing scenarios, explanations, and interpretations, our ability to assess their likelihood, or to evaluate them critically, is grossly inadequate. Once we have adopted a particular hypothesis or interpretation, we grossly exaggerate the likelihood of that hypothe-

sis, and find it very difficult to see things any other way.

Amos was polite about it. He did not say, as he often said, "It is amazing how dull history books are, given how much of what's in them must be invented." What he did say was perhaps even more shocking to his audience: Like other human beings, historians were prone to the cognitive biases that he and Danny had described. "Historical judgment," he said, was "part of a broader class of processes involving intuitive interpretation of data." Historical judgments were subject to bias. As an example, Amos talked about research then being conducted by one of his graduate students at Hebrew University, Baruch Fischhoff. When Richard Nixon announced his surprising intention to visit China and Russia, Fischhoff asked people to assign odds to a list of possible outcomes — say, that Nixon would meet Chairman Mao at least once, that the United States and the Soviet Union would create a joint space program, that a group of Soviet Jews would be arrested for attempting to speak with Nixon, and so on. After the trip, Fischhoff went back and asked the same people to recall the odds they had assigned to each outcome. Their memories of the odds they had assigned to various outcomes were badly distorted. They all believed that they had assigned higher probabilities to what happened than they actually had. They greatly overestimated the odds that they had assigned to what had actually happened. That is, once they knew the outcome, they thought it had been far more predictable than they had found it to be before, when they had tried to predict it. A few years after Amos described the work to his Buffalo audience, Fischhoff named the phenomenon "hindsight bias."*

In his talk to the historians, Amos de-

* In a brief memoir, Fischhoff later recalled how his idea had first come to him in Danny's seminar: "We read Paul Meehl's (1973) 'Why I Do Not Attend Case Conferences.' One of his many insights concerned clinicians' exaggerated feeling of having known all along how cases were going to turn out." The conversation about Meehl's idea led Fischhoff to think about the way Israelis were always pretending to have foreseen essentially unforeseeable political events. Fischhoff thought, "If we're so prescient, why aren't we running the world?" Then he set out to see exactly how prescient people who thought themselves prescient actually were.

scribed their occupational hazard: the tendency to take whatever facts they had observed (neglecting the many facts that they did not or could not observe) and make them fit neatly into a confident-sounding story:

> All too often, we find ourselves unable to predict what will happen; yet after the fact we explain what did happen with a great deal of confidence. This "ability" to explain that which we cannot predict, even in the absence of any additional information, represents an important, though subtle, flaw in our reasoning. It leads us to believe that there is a less uncertain world than there actually is, and that we are less bright than we actually might be. For if we can explain tomorrow what we cannot predict today, without any added information except the knowledge of the actual outcome, then this outcome must have been determined in advance and we should have been able to predict it. The fact that we couldn't is taken as an indication of our limited intelligence rather than of the uncertainty that is in the world. All too often, we feel like kicking ourselves for failing to foresee that which later appears inevitable. For all we know, the handwrit-

ing might have been on the wall all along. The question is: was the ink visible?

It wasn't just sports announcers and political pundits who radically revised their narratives, or shifted focus, so that their stories seemed to fit whatever had just happened in a game or an election. Historians imposed false order upon random events, too, probably without even realizing what they were doing. Amos had a phrase for this. "Creeping determinism," he called it — and jotted in his notes one of its many costs: "He who sees the past as surprise-free is bound to have a future full of surprises."

A false view of what has happened in the past makes it harder to see what might occur in the future. The historians in his audience of course prided themselves on their "ability" to construct, out of fragments of some past reality, explanatory narratives of events which made them seem, in retrospect, almost predictable. The only question that remained, once the historian had explained how and why some event had occurred, was why the people in his narrative had not seen what the historian could now see. "All the historians attended Amos's talk," recalled Biederman, "and they left ashen-faced."

After he had heard Amos explain how the mind arranged historical facts in ways that made past events feel a lot less uncertain, and a lot more predictable, than they actually were, Biederman felt certain that his and Danny's work could infect any discipline in which experts were required to judge the odds of an uncertain situation — which is to say, great swaths of human activity. And yet the ideas that Danny and Amos were generating were still very much confined to academia. Some professors, most of them professors of psychology, had heard of them. And no one else. It was not at all clear how two guys working in relative obscurity at Hebrew University could spread the word of their discoveries to people outside their field.

In the early months of 1973, after their return to Israel from Eugene, Amos and Danny set to work on a long article summarizing their findings. They wanted to gather in one place the chief insights of the four papers they had already written and allow readers to decide what to make of them. "We decided to present the work for what it was: a psychological investigation," said Danny. "We'd leave the big implications to others." He and Amos both agreed that the journal *Science* offered them the best hope of reaching people in fields outside of psychology.

Their article was less written than it was constructed. ("A sentence was a good day," said Danny). As they were building it, they stumbled upon what they saw as a clear path for their ideas to enter everyday human life. They had been gripped by "The Decision to Seed Hurricanes," a paper coauthored by Stanford professor Ron Howard. Howard was one of the founders of a new field called decision analysis. Its idea was to force decision makers to assign probabilities to various outcomes: to make explicit the thinking that went into their decisions before they made them. How to deal with killer hurricanes was one example of a problem that policy makers might use decision analysts to help address. Hurricane Camille had just wiped out a large tract of the Mississippi Gulf Coast and obviously might have done a lot more damage — say, if it had hit New Orleans or Miami. Meteorologists thought they now had a technique — dumping silver iodide into the storm — to reduce the force of a hurricane, and possibly even alter its path. Seeding a hurricane wasn't a simple matter, however. The moment the government intervened in the storm, it was implicated in whatever damage that storm inflicted. The public, and the courts of law, were unlikely to give the government credit for what had *not* happened, for who could say with certainty what would have happened if the government had not intervened? Instead the society would hold its leaders responsible for whatever damage the storm inflicted, wherever it hit. Howard's paper explored how the government might decide what to do — and that involved estimating the odds of various outcomes.

But the way the decision analysts elicited probabilities from the minds of the hurricane experts was, in Danny and Amos's eyes, bizarre. The analysts would present the hurricane seeding experts inside government with a wheel of fortune on which, say, a third of the slots were painted red. They'd ask: "Would you rather bet on the red sector of this wheel or bet that the seeded hurricane will cause more than $30 billion of property damage?" If the hurricane authority said he would rather bet on red, he was saying that he thought the chance the hurricane would cause more than $30 billion of property damage was less than 33 percent. And so the decision analysts would show him another wheel, with, say, 20 percent of the slots painted red. They did this until the percentage of red slots matched up with the authority's sense of the odds that the hurricane would cause more than $30 billion of property damage. They just assumed that the hurricane seeding experts had an ability to correctly assess the odds of highly uncertain events.

Danny and Amos had already shown that people's ability to judge probabilities was queered by various mechanisms used by the mind when it faced uncertainty. They believed that they could use their new understanding of the systematic errors in people's judgment to improve that judgment — and, thus, to improve people's decision making. For instance, any person's assessment of probabilities of a killer storm making landfall in 1973 was bound to be warped by the ease with which they recalled the fresh experience of Hurricane Camille. But how, exactly, was that judgment warped? "We thought decision analysis would conquer the world and we would help," said Danny.

The leading decision analysts were clustered around Ron Howard in Menlo Park, California, at a place called the Stanford Research Institute. In the fall of 1973 Danny and Amos flew to meet with them. But before they could figure out exactly how they were going to bring their ideas about uncertainty into the real world, uncertainty intervened. On October 6, the armies of Egypt and Syria — with troops and planes and money from as many as nine other Arab countries — launched an attack on Israel. Israeli intelligence analysts had dramatically misjudged the odds of an attack of any sort, much less a coordinated one. The army was caught off guard. On the Golan Heights, a hundred or so Israeli tanks faced fourteen hundred Syrian tanks. Along the Suez Canal, a garrison of five hundred Israeli troops and three tanks were quickly overrun by two thousand Egyptian tanks and one hundred thousand Egyptian soldiers. On a cool, cloudless, perfect morning in Menlo Park, Amos and Danny heard the news of the shocking Israeli losses. They raced to the airport for the first flight back home, so that they might fight in yet another war.

8
Going Viral

The young woman they called him to examine that summer day was still in a state of shock. As Don Redelmeier understood it, her car had smashed head-on into another car a few hours earlier, and the ambulance had rushed her straight to Sunnybrook Hospital. She'd suffered broken bones everywhere — some of which they had detected and others, it later became clear, they had not. They'd found the multiple fractures in her ankles, feet, hips, and face. (They'd missed the fractures in her ribs.) But it was only after she arrived in the Sunnybrook operating room that they realized there was something wrong with her heart.

Sunnybrook was Canada's first and largest regional trauma center, an eruption of red-brown bricks in a quiet Toronto suburb. It had started its life as a hospital for soldiers returning from the Second World War, but as the veterans died, its purpose shifted. In the 1960s the government finished building what would become at its widest a twenty-four-lane highway across Ontario. It would also become the most heavily used road in North America, and one of its busiest stretches passed close by the hospital. The carnage from Highway 401 gave the hospital a new life. Sunnybrook rapidly acquired a reputation for treating victims of automobile accidents; its ability to cope with one sort of medical trauma inevitably attracted other sorts of trauma. "Business begets business," explained one of Sunnybrook's administrators. By the turn of the twenty-first century, Sunnybrook was the go-to destination not only for victims of car crashes but for attempted suicides, wounded police officers, old people who had taken a fall, pregnant women with serious complications, construction workers who had been hurt on the job, and the survivors of gruesome snowmobile crashes — who were medevaced in with surprising frequency from the northern Canadian boondocks. Along with the trauma came complexity. A lot of the damaged people who turned up at Sunnybrook had more than one thing wrong with them.

That's where Redelmeier entered. By

nature a generalist, and by training an internist, his job in the trauma center was, in part, to check the understanding of the specialists for mental errors. "It isn't explicit but it's acknowledged that he will serve as a check on other people's thinking," said Rob Fowler, an epidemiologist at Sunnybrook. "About *how* people do their thinking. He keeps people honest. The first time people interact with him they'll be taken aback: Who the hell is this guy, and why is he giving me feedback? But he's lovable, at least the second time you meet him." That Sunnybrook's doctors had come to appreciate the need for a person to serve as a check on their thinking, Redelmeier thought, was a sign of how much the profession had changed since he entered it in the mid-1980s. When he'd started out, doctors set themselves up as infallible experts; now there was a place in Canada's leading regional trauma center for a connoisseur of medical error. A hospital was now viewed not just as a place to treat the unwell but also as a machine for coping with uncertainty. "Wherever there is uncertainty there has got to be judgment," said Redelmeier, "and wherever there is judgment there is an opportunity for human fallibility."

Across North America, more people died

every year as a result of preventable accidents in hospitals than died in car crashes — which was saying something. Bad things happened to patients, Redelmeier often pointed out, when they were moved without extreme care from one place in a hospital to another. Bad things happened when patients were treated by doctors and nurses who had forgotten to wash their hands. Bad things even happened to people when they pressed hospital elevator buttons. Redelmeier had actually co-written an article about that: "Elevator Buttons as Unrecognized Sources of Bacterial Colonization in Hospitals." For one of his studies, he had swabbed 120 elevator buttons and 96 toilet seats at three big Toronto hospitals and produced evidence that the elevator buttons were far more likely to infect you with some disease.

But of all the bad things that happened to people in hospitals, the one that most preoccupied Redelmeier was clinical misjudgment. Doctors and nurses were human, too. They sometimes failed to see that the information patients offered them was unreliable — for instance, patients often said that they were feeling better, and might indeed believe themselves to be improving, when they had experienced no real change in their condition. Doctors tended to pay attention mainly to what they were asked to pay attention to, and to miss some bigger picture. They sometimes failed to notice what they were not directly assigned to notice. "One of the things Don taught me was the value of observing the room when the patient isn't there," says Jon Zipursky, chief of residents at Sunnybrook. "Look at their meal tray. Did they eat? Did they pack for a long stay or a short one? Is the room messy or neat? Once we walked into the room and the patient was sleeping. I was about to wake him up and Don stops me and says, *There is a lot you can learn about people from just watching.*"

Doctors tended to see only what they were trained to see: That was another big reason bad things might happen to a patient inside a hospital. A patient received treatment for something that was obviously wrong with him, from a specialist oblivious to the possibility that some less obvious thing might also be wrong with him. The less obvious thing, on occasion, could kill a person.

The conditions of people mangled on the 401 were often so dire that the most obvious things wrong with them demanded the complete attention of the medical staff, and immediate treatment. But the dazed young woman who arrived in the Sunnybrook emergency room directly from her head-on car crash, with her many broken bones, presented her surgeons, as they treated her, with a disturbing problem. The rhythm of her heartbeat had become wildly irregular. It was either skipping beats or adding extra beats; in any case, she had more than one thing seriously wrong with her.

Immediately after the trauma center staff called Redelmeier to come to the operating room, they diagnosed the heart problem on their own — or thought they had. The young woman remained alert enough to tell them that she had a past history of an overactive thyroid. An overactive thyroid can cause an irregular heartbeat. And so, when Redelmeier arrived, the staff no longer needed him to investigate the source of the irregular heartbeat but to treat it. No one in the operating room would have batted an eye if Redelmeier had simply administered the drugs for hyperthyroidism. Instead, Redelmeier asked everyone to slow down. To wait. Just a moment. Just to check their thinking — and to make sure they were not trying to force the facts into an easy, coherent, but ultimately false story.

Something bothered him. As he said later, "Hyperthyroidism is a classic cause of an irregular heart rhythm, but hyperthyroidism is an *infrequent* cause of an irregular heart rhythm." Hearing that the young woman had a history of excess thyroid hormone production, the emergency room medical staff had leaped, with seeming reason, to the assumption that her overactive thyroid had caused the dangerous beating of her heart. They hadn't bothered to consider statistically far more likely causes of an irregular heartbeat. In Redelmeier's experience, doctors did not think statistically. "Eighty percent of doctors don't think probabilities apply to their patients," he said. "Just like 95 percent of married couples don't believe the 50 percent divorce rate applies to them, and 95 percent of drunk drivers don't think the statistics that show that you are more likely to be killed if you are driving drunk than if you are driving sober applies to them."

Redelmeier asked the emergency room staff to search for other, more statistically likely causes of the woman's irregular heartbeat. That's when they found her collapsed lung. Like her fractured ribs, her collapsed lung had failed to turn up on the X-ray. Unlike the fractured ribs, it could kill her. Redelmeier ignored the thyroid and treated the collapsed lung. The young woman's heartbeat returned to normal. The next day, her formal thyroid tests came back: Her thyroid hormone production was perfectly normal. Her thyroid never had been the issue. "It was a classic case of the representativeness heuristic," said Redelmeier. "You need to be so careful when there is one simple diagnosis that instantly pops into your mind that beautifully explains everything all at once. That's when you need to stop and check your thinking."

It wasn't that what first came to mind was always wrong; it was that its existence in your mind led you to feel more certain than you should be that it was correct. "Beware of the delirious guy in the emergency unit with the long history of alcoholism," said Redelmeier, "because you will say, 'He's just drunk,' and you'll miss the subdural hematoma." The woman's surgeons had leapt from her medical history to a diagnosis without considering the base rates. As Kahneman and Tversky long ago had pointed out, a person who is making a prediction — or a diagnosis — is allowed to ignore base rates only if he is completely certain he is correct. Inside a hospital, or really anyplace else, Redelmeier was never completely certain about anything, and he didn't see why anybody else should be, either.

■ ■ ■ ■

Redelmeier had grown up in Toronto, in the same house in which his stockbroker father had been raised. The youngest of three boys, he often felt a little stupid; his older brothers always seemed to know more than he did and were keen to let him know it. Redelmeier also had a speech impediment — a maddening stammer he would never cease to work hard, and painfully, to compensate for. (When he called for restaurant reservations, he just told them his name was "Don Red.") His stammer slowed him down when he spoke; his weakness as a speller slowed him down when he wrote. His body was not terribly well coordinated, and by the fifth grade he required glasses to correct his eyesight. His two great strengths were his mind and his temperament. He was always extremely good at math; he loved math. He could explain it, too, and other kids came to him when they couldn't understand what the teacher had said. That is where his temperament entered. He was almost peculiarly considerate of others. From the time he was a small child, grown-ups had noticed that about him: His first instinct upon meeting someone else was to take care of the person.

Still, even from math class, where he often wound up helping all the other students, what he took away was a sense of his own fallibility. In math there was a right answer and a wrong answer, and you couldn't fudge it. "And the errors are sometimes predictable," he said. "You see them coming a mile away and you *still* make them." His experience of life as an error-filled sequence of events, he later thought, might be what had made him so receptive to an obscure article, in the journal *Science,* that his favorite high school teacher, Mr. Fleming, had given him to read in late 1977. He took the article home with him and read it that night at his desk.

The article was called "Judgment Under Uncertainty: Heuristics and Biases." It was in equal parts familiar and strange — what the hell was a "heuristic"? Redelmeier was seventeen years old, and some of the jargon was beyond him. But the article described three ways in which people made judgments when they didn't know the answer for sure. The names the authors had given these — representativeness, availability, anchoring — were at once weird and seductive. They made the phenomenon they described feel like secret knowledge. And yet what they were saying struck Redelmeier as the simple truth — mainly because he was fooled by the questions they put to the reader. He, too, guessed that the guy they named "Dick" and described so blandly was equally likely to be a lawyer or an engineer, even though he came from a pool that was mostly lawyers. He, too, made a different prediction when he was given worthless evidence than when he was given no evidence at all. He, too, thought that there were more words in a typical passage of English prose that started with *K* than had *K* in the third position, because the words that began with *K* were easier to recall. He, too, made predictions about people from mere descriptions of them with a degree of confidence that was totally unjustified — even uncertain Don Redelmeier fell prey to overconfidence! And when asked quickly to guess the product of 1 × 2 × 3 × 4 × 5 × 6 × 7 × 8, he saw how he, too, would think it less than the product of 8 × 7 × 6 × 5 × 4 × 3 × 2 × 1.

What struck Redelmeier wasn't the idea that people made mistakes. Of course people made mistakes! What was so compelling is that the mistakes were predictable and systematic. They seemed ingrained in human nature. Reading the article in *Sci-*

ence reminded Redelmeier of all the times he had made what seemed in retrospect to be an obvious mistake on a math problem — because it was so much like the other mistakes he and others had made. One passage in particular stuck with him — it was in the section on this thing they called "availability." It talked about the role of the imagination in human error. "The risk involved in an adventurous expedition, for example, is evaluated by imagining contingencies with which the expedition is not equipped to cope," the authors wrote. "If many such difficulties are vividly portrayed, the expedition can be made to appear exceedingly dangerous, although the ease with which disasters are imagined need not reflect their actual likelihood. Conversely, the risk involved in an undertaking may be grossly underestimated if some possible dangers are either difficult to conceive of, or simply do not come to mind."

This wasn't just about how many words in the English language started with the letter *K*. This was about life and death. "That article was more thrilling than a movie to me," said Redelmeier. "And I love movies."

Redelmeier had never heard of the authors — Daniel Kahneman and Amos Tversky — though at the bottom of the page it said that they were members of the Department of Psychology at Hebrew University in Jerusalem. To him it was more important that his older brothers had never heard of them, either. *Aha, finally. I know something more than my brothers!* he thought. Kahneman and Tversky offered what felt like a private glimpse of the act of thinking. Reading their article was like getting a peek behind the magician's curtain.

Redelmeier didn't have much trouble figuring out what he wanted to do with his life. As a kid he'd fallen in love with the doctors on television — Leonard McCoy on *Star Trek* and, especially, Hawkeye Pierce on *M*A*S*H.* "I sort of wanted to be heroic," he said. "I would never cut it in sports. I would never cut it in politics. I would never make it in the movies. Medicine was a path. A way to have a truly heroic life." He felt the pull so strongly that he applied to medical school at the age of nineteen, during his second year of college. Just after his twentieth birthday he was training, at the University of Toronto, to become a doctor.

And that's where the problems started: The professors didn't have much in common with Leonard McCoy or Hawkeye Pierce. A lot of them were self-important and even a bit pompous. Something about them, and what they were saying, led Redelmeier to seditious thoughts. "Early on in medical school there are a whole bunch of professors who are saying things that are wrong," he recalled. "I don't dare say anything about it." They repeated common superstitions as if they were eternal truths. ("Bad things come in threes.") Specialists from different fields of medicine faced with the same disease offered contradictory diagnoses. His professor of urology told students that blood in the urine suggested a high chance of kidney cancer, while his professor of nephrology said that blood in the urine indicated a high chance of glomerulonephritis — kidney inflammation. "Both had exaggerated confidence based on their expert experience," said Redelmeier, and both mainly saw only what they had been trained to see.

The problem was not what they knew, or didn't know. It was their need for certainty or, at least, the appearance of certainty. Standing beside the slide projector, many of them did not so much teach as preach. "There was a generalized mood of arrogance," said Redelmeier. " *'What do you mean you didn't give steroids!!????'* " To Redelmeier the very idea that there was a great deal of uncertainty in medicine went largely unacknowledged by its authorities.

There was a reason for this: To acknowledge uncertainty was to admit the possibility of error. The entire profession had arranged itself as if to confirm the wisdom of its decisions. Whenever a patient recovered, for instance, the doctor typically attributed the recovery to the treatment he had prescribed, without any solid evidence that the treatment was responsible. *Just because the patient is better after I treated him doesn't mean he got better because I treated him,* Redelmeier thought. "So many diseases are self-limiting," he said. "They will cure themselves. People who are in distress seek care. When they seek care, physicians feel the need to do something. You put leeches on; the condition improves. And that can propel a lifetime of leeches. A lifetime of overprescribing antibiotics. A lifetime of giving tonsillectomies to people with ear infections. You try it and they get better the next day and it is so compelling. You go to see a psychiatrist and your depression improves — you are convinced of the efficacy of psychiatry."

Redelmeier noticed other problems, too. His medical school professors took data at face value that should have been inspected more closely, for example. An old man would come into the hospital suffering from pneumonia. They'd check his heart rate and find it to be a reassuringly normal seventy-five beats per minute . . . and just move on. But the reason pneumonia killed so many old people was its power to spread infection. An immune system responding as it should generated fever, coughs, chills, sputum — and a faster than normal heartbeat. A body fighting an infection required blood to be pumped through it at a faster than normal rate. "The heart rate of an old man with pneumonia is not supposed to be normal!" said Redelmeier. "It's supposed to be ripping along!" An old man with pneumonia whose heart rate appears normal is an old man whose heart may well have a serious problem. But the normal reading on the heart rate monitor created a false sense in doctors' minds that all was well. And it was precisely when all seemed well that medical experts "failed to check themselves."

As it happens, a movement was taking shape right then and there in Toronto that came to be called "evidence-based medicine." The core idea of evidence-based medicine was to test the intuition of medical experts — to check the thinking of doctors against hard data. When subjected to scientific investigation, some of what passed for medical wisdom turned out to be shockingly wrong-headed. When Redelmeier entered medical school in 1980, for instance, the conventional wisdom held that if a heart attack victim suffered from some subsequent arrhythmia, you gave him drugs to suppress it. By the end of Redelmeier's medical training, seven years later, researchers had shown that heart attack patients whose arrhythmia was suppressed died more often than the ones whose condition went untreated. No one explained why doctors, for years, had opted for a treatment that systematically killed patients — though proponents of evidence-based medicine were beginning to look to the work of Kahneman and Tversky for possible explanations. But it was clear that the intuitive judgments of doctors could be gravely flawed: The evidence of the medical trials now could not be ignored. And Redelmeier was alive to the evidence. "I became very aware of the buried analysis — that a lot of the probabilities were being made up by expert opinion," said Redelmeier. "I saw error in the way people think that was being transmitted to patients. And people had no recognition of the mistakes that they were making. I had a little unhappiness, a little dissatisfaction, a sense that all was not right in the state of Denmark."

Toward the end of their article in *Science,* Daniel Kahneman and Amos Tversky had pointed out that, while statistically sophisticated people might avoid the simple mistakes made by less savvy people, even the most sophisticated minds were prone to error. As they put it, "their intuitive judgments are liable to similar fallacies in more intricate and less transparent problems." That, the young Redelmeier realized, was a "fantastic rationale why brilliant physicians were not immune to these fallibilities." He thought back to the errors he had made while trying to solve math problems. "The same problem solving exists in medicine," he said. "In math you always check your work. In medicine, no. And if we are fallible in algebra, where the answers are clear, how much more fallible must we be in a world where the answers are much less clear?" Error wasn't necessarily shameful; it was merely human. "They provided a language and a logic for articulating some of the pitfalls people encounter when they think. Now these mistakes could be communicated. It was the recognition of human error. Not its denial. Not its demonization. Just the understanding that they are part of human nature."

But Redelmeier kept to himself any heretical thoughts he harbored as a young medical student. He had never felt the impulse to question authority or flout convention, and had no talent for either. "I was never shocked and disappointed before in my life," he said. "I was always very obedient. Law-abiding. I vote in all elections. I show up at every university staff meeting. I've never had an altercation with the police."

In 1985, he was accepted as a medical resident at the Stanford University hospital. At Stanford he began, haltingly, to voice his professional skepticism. One night during his second year, he was manning the intensive care unit and was assigned to keep a young man alive long enough to harvest his organs. (The American euphemism — "harvesting" — sounded strange to his ears. In Canada they called it "organ retrieval.") His brain-dead patient was a twenty-one-year-old who had wrapped his motorcycle around a tree.

It was the first time Redelmeier had been confronted with the dying body of a person younger than himself, and it bothered him, in a way that the deaths of older people he had witnessed had not. "It was such a loss of *so* many life years," he said. "It was such a preventable case. And the guy hadn't been wearing a helmet." Redelmeier was newly struck by the inability of human beings to judge risks, even when their misjudgment might kill them. When making judgments, people obviously could use help — say, by requiring all motorcyclists to wear helmets. Later Redelmeier said as much to one of his fellow students, an American. *What is it with you freedom-loving Americans?* he asked. *Live free or die. I don't get it. I say, "Regulate me gently. I'd rather live."* His fellow student replied, *Not only do a lot of Americans not share your view; other physicians don't share your view.* Redelmeier's fellow student told him about Stanford's famous head of cardiac surgery, Norm Shumway, who had actively lobbied against the creation of a law that would require motorcyclists to wear helmets. "It dropped my jaw," said Redelmeier. How could a guy so smart be so stupid about that? We're definitely capable of errors. And human fallibility should be paid attention to."

At the age of twenty-seven, as he finished his Stanford residency, Redelmeier was creating the beginnings of a worldview that internalized the article by the two Israeli psychologists that he had read as a teenager. Where this worldview would lead he did not know. He still thought it possible that, upon his return to Canada, he might just move back up to northern Labrador, where he had spent one summer during medical school delivering health care to a village of five hundred people. "I didn't have great memory skills or great dexterity," he said. "I was afraid I wouldn't be a great doctor. And if I wasn't going to be great, I might as well go to serve someplace that was underserved, where I was needed and wanted." Redelmeier actually still believed that he might wind up practicing medicine in a conventional manner. But then he met Amos Tversky.

Redelmeier had long made a habit of anticipating his own mental errors and correcting for them. Alive to the fallibility of his memory, he carried a notepad wherever he went and wrote down thoughts and problems as they occurred to him. When awakened late at night by a phone call from the hospital, he always lied and told the fast-talking resident on the other end of the line that they had a bad connection, and so he needed to repeat everything he had just said. "You can't tell a resident he is speaking too quickly. You blame yourself — and it facilitates not only his thinking but my own." When a visitor turned up in Redelmeier's office when he was between rounds, he would set a kitchen timer to make sure he didn't get lost in conversation and wind up late for his patients. "Redelmeier loses track of time when he is having fun," said Redelmeier. In advance of any social situation, he went to unusual lengths to correct whatever he imagined might go wrong. When he gave a talk — still a massive challenge for him, with his stammer — he cased the lecture hall and simulated his entire performance.

And so, in the spring of 1988, for Redelmeier it felt perfectly normal, two days before his first lunch with Amos Tversky, to walk through the Stanford Faculty Club dining room where they were scheduled to meet. On the day of the lunch, he moved his hospital tour of patients from 6:30 in the morning to 4:30, to reduce the risk that anyone's medical problems would interfere with his meeting. He didn't eat breakfast usually, but on this day he did, so that he wouldn't be distracted by hunger during lunch. As was also his habit, he jotted down in advance little notes — potential topics of discussion — "for fear of blanking." Not that he intended to say much. Hal Sox, Redelmeier's superior at Stanford, who would be joining them, had told Redelmeier, "Don't talk. Don't say anything. Don't interrupt. Just sit and listen." Meeting with Amos Tversky, Hal Sox, said, was "like brainstorming with Albert Einstein. He is one for the ages — there won't ever be anyone else like him."

Hal Sox happened to have coauthored the first article Amos ever wrote about medicine. Their paper had sprung from a question Amos had put to Sox: How did a tendency people exhibited when faced with financial gambles play itself out in the minds of doctors and patients? Specifically, given a choice between a sure gain and a bet with the same expected value (say, $100 for sure or a 50-50 shot at winning $200), Amos had explained to Hal Sox, people tended to take the sure thing. A bird in the hand. But, given the choice between a sure *loss* of $100 and a 50-50 shot of losing $200, they took the risk. With Amos's help, Sox and two other medical researchers designed experiments to show how differently both doctors and patients made choices when those choices were framed in terms of losses rather than gains.

Lung cancer proved to be a handy example. Lung cancer doctors and patients in the early 1980s faced two unequally unpleasant options: surgery or radiation. Surgery was more likely to extend your life, but, unlike radiation, it came with the small risk of instant death. When you told people that they had a 90 percent chance of surviving surgery, 82 percent of patients opted for surgery. But when you told them that they had a 10 percent chance of *dying* from the surgery — which was of course just a different way of putting the same odds — only 54 percent chose the surgery. People facing a life-and-death decision responded not to the odds but to the way the odds were described to them. And not just patients; doctors did it, too. Working with Amos, Sox said, had altered his view of his own profession. "The cognitive aspects are not at all understood in medicine," he said. Among other things, he could not help but wonder how many surgeons, consciously or unconsciously, had told some patient that he had a 90 percent chance of surviving a surgery, rather than a 10 percent of dying from it, simply because it was in his interest to perform the surgery.

At that first lunch, Redelmeier mainly just watched as Sox and Amos talked. Still, he noticed some things. Amos's pale blue eyes darted around, and he had a slight speech impediment. His English was fluent but spoken with a thick Israeli accent. "He was a little bit hypervigilant," said Redelmeier. "He was bouncy. Energetic. He had none of the lassitude of some of the tenured faculty. He did 90 percent of the talking. Every word was worth listening to. I was surprised by how little medicine he knew, because he was already having a big effect on medical decision making." Amos had all sorts of questions for the two doctors; most of them had to do with probing for illogic in medical behavior. After watching Hal Sox answer or try to answer Amos's questions, Redelmeier realized that he was learning more about his superior in a single lunch than he'd gathered from the previous three years. "Amos knew exactly what questions to ask," said Redelmeier. "There were no awkward silences."

At the end of the lunch, Amos invited Redelmeier to visit him in his office. It didn't take long before Amos was bouncing ideas about the human mind off Redelmeier, as he had bounced them off Hal Sox, to listen for an echo in medicine. The Samuelson bet, for instance. The Samuelson bet was named for Paul Samuelson, the economist who had cooked it up. As Amos explained it, people offered a single bet in which they have a 50-50 chance either to win $150 or lose $100 usually decline it. But if you offer those same people the chance to make the same bet one hundred times over, most of them accept the bet. Why did they make the expected value calculation — and respond to the odds being in their favor — when they were allowed to make the bet a hundred times, but not when they are offered a single bet? The answer was not entirely obvious. Yes, the more times you play a game with the odds in your favor, the less likely you are to lose; but the more times you play, the greater the total sum of money you stood to lose. Anyway, after Amos finishing explaining the paradox, "He said, 'Okay, Redelmeier, find me the medical analogy to that!' "

For Redelmeier, medical analogies popped quickly to mind. "Whatever the general example was, I knew a bunch of instantaneous medical examples. It was just astonishing that he would shut up and listen to me." A medical analogy of Samuelson's bet, Redelmeier decided, could be found in the duality in the role of the physician. "The physician is meant to be the perfect agent for the patient as well as the protector of society," he said. "Physicians deal with patients one at a time, whereas health policy makers deal with aggregates."

But there was a conflict between the two roles. The safest treatment for any one patient, for instance, might be a course of antibiotics; but the larger society suffers when antibiotics are overprescribed and the bacteria they were meant to treat evolved into versions of themselves that were more dangerous and difficult to treat. A doctor who did his job properly really could not just consider the interests of the individual patient; he needed to consider the aggregate of patients with that illness. The issue was even bigger than one of public health policy. Doctors saw the same illness over and again. Treating patients, they weren't merely making a single bet; they were being asked to make that same bet over and over again. Did doctors behave differently when they were offered a single gamble and when they were offered the same gamble repeatedly?

The paper subsequently written by Amos with Redelmeier* showed that, in treating individual patients, the doctors behaved differently than they did when they designed ideal treatments for groups of patients with the same symptoms. They were likely to

* "Discrepancy between Medical Decisions for Individual Patients and for Groups" appeared in the *New England Journal of Medicine* in April 1990.

order additional tests to avoid raising troubling issues, and less likely to ask if patients wished to donate their organs if they died. In treating individual patients, doctors often did things they would disapprove of if they were creating a public policy to treat groups of patients with the exact same illness. Doctors all agreed that, if required by law, they should report the names of patients diagnosed with a seizure disorder, diabetes, or some other condition that might lead to loss of consciousness while driving a car. In practice, they didn't do this — which could hardly be in the interest even of the individual patient in question. "This result is not just another manifestation of the conflict between the interests of the patient and the general interests of society," Tversky and Redelmeier wrote, in a letter to the editor of the *New England Journal of Medicine.* "The discrepancy between the aggregate and the individual perspectives also exists in the mind of the physician. The discrepancy seems to call for a resolution; it is odd to endorse a treatment in every case and reject it in general, or vice versa."

The point was not that the doctor was incorrectly or inadequately treating individual patients. The point was that he could not treat his patient one way, and groups of patients suffering from precisely the same problem in another way, and be doing his best in both cases. Both could not be right. And the point was obviously troubling — at least to the doctors who flooded the *New England Journal of Medicine* with letters written in response to the article. "Most physicians try to maintain this facade of being rational and scientific and logical and it's a great lie," said Redelmeier. "A partial lie. What leads us is hopes and dreams and emotion."

Redelmeier's first article with Amos led to other ideas. Soon they were meeting not in Amos's office in the afternoon but at Amos's home late at night. Working with Amos wasn't work. "It was pure joy," said Redelmeier. "Pure fun." Redelmeier knew at some deep level that he was in the presence of a person who would change his life. Many sentences popped out of Amos's mouth that Redelmeier knew he would forever remember:

A part of good science is to see what everyone else can see but think what no one else has ever said.

The difference between being very smart and very foolish is often very small.

So many problems occur when people fail to be obedient when they are supposed to be obedient, and fail to be creative when they are supposed to be creative.

The secret to doing good research is always to be a little underemployed. You waste years by not being able to waste hours.

It is sometimes easier to make the world a better place than to prove you have made the world a better place.

Redelmeier half suspected that the reason Amos had so much time for him was that Redelmeier was not married, and was willing to treat the hours between midnight and four in the morning as part of a workday. The hours Amos kept were strange, but the discipline he imposed became familiar. "He needs the concrete examples to test his general theories," said Redelmeier. "Some of the principles were just extremely robust, and I was supposed to find examples and give voice to them in a particular domain, medicine." Amos had a clear idea of how people misperceived randomness, for instance. They didn't understand that random sequences seemed to have patterns in them: People had incredible ability to see meaning in these patterns where none existed. Watch any NBA game, Amos explained to Redelmeier, and you saw that the announcers, the fans, and maybe even the coaches seemed to believe that basketball shooters had the "hot hand." Simply because some player had made his last few shots, he was thought to be more likely to make his next shot. Amos had collected data on NBA shooting streaks to see if the so-called hot hand was statistically significant — he already could persuade you that it was not. A better shooter was of course more likely to make his next shot than a less able shooter, but the streaks observed by fans and announcers and the players themselves were illusions. He asked Redelmeier to find in medicine the same sort of false pattern–seeking behavior exhibited by basketball announcers.

Redelmeier soon returned with the widely held belief that arthritis pain was related to the weather. For thousands of years, people had imagined this connection; it could be traced back to Hippocrates, who wrote, in 400 BC, about the effect of wind and rain on disease. In the late 1980s, doctors were still suggesting to arthritis patients that they move to warmer climates. Working with Amos, Redelmeier found a large group of arthritis patients and asked them to report their pain levels. He then matched these to weather reports. Pretty quickly, he and Amos established that, despite the patients' claims that their pain changed with the weather, there was no meaningful correlation between the two. They didn't stop there, however. Amos wanted to explain *why* people saw this connection between their pain and the weather. Redelmeier interviewed the patients whose pain he had proven to be unrelated to the weather: All but one of them still insisted that their pain was related to the weather and cited, as evidence, the few random moments that justified their belief. Basketball experts seized on random streaks as patterns in players' shooting that didn't exist. Arthritis sufferers found patterns in suffering that didn't exist. "We attribute this phenomenon to selective matching," Tversky and Redelmeier wrote.* ". . . For arthritis, selective matching leads people to look for changes in the weather when they experience increased pain, and pay little attention to the weather when their pain is stable. . . . [A] single day of severe pain and extreme

* "On the Belief That Arthritis Pain Is Related to the Weather" appeared in the *Proceedings of the National Academy of Sciences* in April 1996.

weather might sustain a lifetime of belief in a relation between them."

There might not be a pattern in arthritis pain, but, to Redelmeier's eye, there appeared to be a very clear pattern in his collaboration with Amos. Amos had all these general ideas about the pitfalls in the human mind when it was required to render judgments in conditions of uncertainty. Their implications for medicine had gone pretty much entirely unexplored. "Sometimes I felt Amos was pilot-testing ideas in front of me," said Redelmeier. "To see if they were germane to the real world." Redelmeier could not help but sense that medicine, for Amos, was "just the tiniest little sliver of his interests." Another human activity in which to explore the specific consequences of the general ideas he had hatched with Danny Kahneman.

Then Danny himself appeared. In late 1988 or maybe early 1989, Amos introduced them in his office. Danny followed up with a phone call to Redelmeier, in which he said how he, too, might like to explore how doctors and patients made decisions. It turned out that Danny had his own ideas, with their own implications. "When he calls me, Danny is working alone," said Redelmeier. "He wants to introduce another heuristic.

One that is all his own, separate from Amos. The introduction of a fourth heuristic. Because there can't be just three."

One day in the summer of 1982 Danny, in his third year as a professor at the University of British Columbia, had walked into his lab and surprised his graduate students with an announcement: They'd now study happiness. Danny had always been curious about people's ability, or inability, to predict their feelings about their own experiences. Now he wanted to study it. Specifically, he wanted to explore the gap — he had sensed it in himself — between a person's intuitions about what made him happy and what actually made him happy. He thought he might start by having people guess how happy it would make them to come into the lab every day for a week and do something that they said they enjoyed — eat a bowl of ice cream, say, or listen to their favorite song. He might then compare the pleasure they anticipated to the pleasure they experienced, and further compare the pleasure they experienced to the pleasure they remembered. There was clearly a difference to be explored, he argued. At the moment your favorite soccer team wins the World Cup, you are beyond elated; six months later, it means next to nothing to you, really. "For a long time there were no subjects involved," recalled Dale Miller, a graduate student of Danny's. "He was just designing these experiments." What Danny imagined is that people wouldn't be especially good at predicting their own happiness — and his first experiments, on a handful of subjects, suggested he was onto something. A man whom no one would ever have described as happy was setting out, to the wonder of those who knew him, to discover the rules of happiness.

Or maybe he was merely sowing doubt in the minds of people who thought they knew what it meant to be happy. At any rate, by the time Amos introduced him to Redelmeier, Danny had moved from the University of British Columbia to the University of California, Berkeley, and from happiness to unhappiness. He was now investigating not only the gap between people's anticipation of pleasure and their experience of pleasure but also the gap between people's experience of pain and their memory of it. What did it mean if people's prediction of the misery that might be caused by some event was different from the misery they actually experienced when the event occurred, or if people's memory of an experience turned out to be meaningfully different from the experience as it had actually played out? A lot, thought Danny. People had a miserable time for most of their vacation and then returned home and remembered it fondly; people enjoyed a wonderful romance but, because it ended badly, looked back on it mainly with bitterness. They didn't simply experience fixed levels of happiness or unhappiness. They experienced one thing and remembered something else.

When he met Redelmeier, Danny was already running experiments on unhappiness in his Berkeley lab. He'd stick the bare arms of his subjects into buckets of ice water. Each subject was given two painful experiences. He'd then be asked which of the two experiences he'd most like to repeat. Funny things happened when you did this with people. Their memory of pain was different from their experience of it. They remembered moments of maximum pain, and they remembered, especially, how they felt the moment the pain ended. But they didn't particularly remember the length of the painful experience. If you stuck people's arms in ice buckets for three minutes but warmed the water just a bit for another minute or so before allowing them to flee the lab, they remembered the experience more fondly than if you stuck their arms in the bucket for three minutes and removed them at a moment of maximum misery. If you asked them to choose one experiment to repeat, they'd take the first session. That is, people preferred to endure more total pain so long as the experience ended on a more pleasant note.

Danny wanted Redelmeier to find him a real-world medical example of what he was calling the "peak-end rule." It didn't take long for Redelmeier to come up with a bunch, but they settled on colonoscopies. In the late 1980s, colonoscopies were painful, and not merely dreaded. The discomfort of the procedure dissuaded people from returning for another one. By 1990, colon cancer was killing sixty thousand people every year in the United States alone. Many of its victims would have survived had their cancer been detected at an early stage. One of the big reasons colon cancer went undetected was that people found their first colonoscopy so unpleasant that they elected not to return for a second one. Was it possible to alter their memory of the experience so that they might forget how unpleasant it was?

To answer the question, Redelmeier ran an experiment on roughly seven hundred people over a period of a year. One group of patients had the colonoscope yanked out of their rear ends at the end of their colonoscopy without ceremony; the other group felt the tip of the scope lingering in their rectums for an extra three minutes. Those extra three minutes were not pleasant. They were merely less unpleasant than the other procedure. The patients in the first group were on the receiving end of an old-fashioned wham-bam-thank-you-ma'am colonoscopy; those in the second group enjoyed a sweeter, or less painful, ending. The sum total of pain experienced by the second group was, however, greater. The patients in the second group experienced all the pain that those in the first group experienced, plus the extra three minutes' worth.

An hour after the procedure, the researchers entered the recovery room and asked the patients to rate their experience. Those who had been given the less unhappy ending remembered less pain than did the patients who had not. More interestingly, they proved more likely to return for another colonoscopy when the time came. Human beings who had never imagined that they might prefer more pain to less could nearly all be fooled into doing so. As Redelmeier put it, "Last impressions can be lasting impressions."

▪ ▪ ▪ ▪

Working with Danny was different from working with Amos. Redelmeier's mental picture of Amos was always crystal clear. Danny left behind a more complicated and murkier impression. Danny was not joyful: Danny was maybe even depressed. He suffered for his work, and so those who worked with him inevitably suffered a bit, too. "He was more likely to see what was wrong with the work and less likely to see what was right with it," said Redelmeier. And yet what came out of his mind was also, obviously, spectacular.

It was odd, when Redelmeier stopped to think about it, how little he ever learned about Amos's and Danny's lives. "Amos told me very, very little about his life," he said. "He never talked about Israel. He never talked about the wars. He didn't talk about the past. It's not that he was deliberately evasive. It's just that he controlled the agenda." The agenda, when they were together, was to analyze human behavior in the delivery of health care. He didn't presume to ask Danny or Amos about their past or their relationship to each other. And so he never found out how or why they had left Hebrew University and Israel for North America. Or why Amos had spent the 1980s as an exalted chaired professor of behavioral science at Stanford, while Danny passed most of that time in relative obscurity at the University of British Columbia. The two men seemed friendly enough, but they weren't obviously working together: Why was that? Redelmeier didn't know. "And they wouldn't talk about each other," he said.

Instead they seemed to have decided they'd bag more game if they hunted separately rather than together. Both were engaged, in different ways, in extending the ideas that they had given birth to jointly in the real world. "I was thinking they were just buddies and I was their pet schnauzer," said Redelmeier.

Redelmeier returned to Toronto in 1992. The experience working with Amos had been life-altering. The man was so vivid that you could not confront any question without wondering how he would approach it. And yet, as Amos always seemed to have all the big ideas, and simply needed medical examples to illustrate them, Redelmeier was left with the feeling that maybe he hadn't done very much. "In many ways I was a glorified secretary, and that troubled me for many years," he said. "Deep down, I thought I was extremely replaceable. When I came back to Toronto, I wondered: Was it just Amos? Or was there something Redelmeier?"

Still, only a few years earlier, he'd imagined that he might wind up a general practitioner in a small village in northern Labrador. Now he had a particular ambition: to explore, as both researcher and doctor, the mental mistakes that doctors and their patients made. He wanted to combine cognitive psychology, as practiced by Danny and Amos, with medical decision making. How exactly he would do this he could not immediately say. He was still too unsure of himself. All he knew for sure was that by working with Amos Tversky, he had discovered this other side to himself: a seeker of truth. He wanted to use data to find true patterns in human behavior, to replace the false ones that governed people's lives and, often, their deaths. "I didn't really know it was in there," Redelmeier said of this side of himself. "Amos doesn't uncover it. He implants it. He sends me as a messenger to a land in the future that he will never see."

9
Birth of the Warrior Psychologist

By the fall of 1973 it was fairly clear to Danny that other people would never fully understand his relationship with Amos. The previous academic year, they'd taught a seminar together at Hebrew University. From Danny's point of view, it had been a disaster. The warmth he felt when he was alone with Amos vanished whenever Amos was in the presence of an audience. "When we were with other people we were one of two ways," said Danny. "Either we finished each other's sentences and told each other's jokes. Or we were competing. No one ever saw us working together. No one knows what we were like." What they were like, in every way but sexually, was lovers. They connected with each other more deeply than either had connected with anyone else. Their wives noticed it. "Their relationship was more intense than a marriage," said Barbara. "I think they were both turned on intellectually more than either had ever been before. It was as if they were both waiting for it." Danny sensed that his wife felt some jealousy; Amos actually praised Barbara, behind her back, for dealing so gracefully with the intrusion on their marriage. "Just to be with him," said Danny. "I never felt that way with anyone else, really. You are in love and things. But I was *rapt.* And that's what it was like. It was truly extraordinary."

And yet it was Amos who worked hardest to find ways to keep them together. "I was the one who was holding back," said Danny. "I kept my distance because I was afraid of what would happen to me without him."

It was four in the morning California time when the armies of Egypt and Syria launched their attack upon Israel. They'd taken the Israelis by surprise on Yom Kippur. Along the Suez Canal, the 500-man Israeli garrison was overwhelmed by 100,000 Egyptian troops. From the Golan Heights, 177 Israeli tank crews gazed down upon an attacking force of 2,000 Syrian tanks. Amos and Danny, still in the United States trying to become decision analysts, raced to the airport and got the first flight possible to Paris, where Danny's sister worked in the Israeli embassy. Getting into

Israel during a war wasn't easy. Every inbound El Al plane was crammed with fighter pilots and combat unit commanders who were coming in to replace the men killed in the first days of the invasion. That's just what you did, if you were an Israeli capable of fighting in 1973: You ran toward the war. Knowing this, Egyptian President Anwar Sadat had promised to shoot down any commercial planes attempting to land in Israel. As they waited in Paris for Danny's sister to talk someone into letting them onto a flight, Danny and Amos bought combat boots. They were made of canvas — lighter than the leather boots issued by the Israeli military.

When the war broke out, Barbara Tversky was on the way to an emergency room in Jerusalem with her eldest son. He had won a contest with his brother to see who could stick a cucumber farthest up his own nose. As they headed home, people surrounded their car and screamed at Barbara for being on the road. The country was in a state of panic: Fighter jets screamed low over Jerusalem to signal all reserves to return to their units. Hebrew University closed. Army trucks rumbled all night through the Tverskys' usually tranquil neighborhood. The city was black. Street lamps remained off;

anyone who owned a car taped over its brake lights. The stars could not have been more spectacular, or the news more troubling — because, for the first time, Barbara sensed that the Israeli government was withholding the truth. This war was different from the others: Israel was losing. Not knowing where Amos was, or what he planned to do, didn't help. Phone calls were so expensive that when he was in the United States they communicated only by letter. Her situation wasn't unusual: There were Israelis who would learn that loved ones living abroad had returned to Israel to fight only by being informed that they had been killed in action.

To make herself useful, Barbara went to the library and found the material to write a newspaper article about stress and how to cope with it. A few nights into the conflict, around ten o'clock, she heard footsteps. She was working alone in the study, with the blinds lowered, to avoid letting the light seep out. The kids were asleep. Whoever was coming up the stairs was running; then suddenly Amos bounded from the darkness. The El Al flight that he had taken with Danny had carried as passengers no one but Israeli men returning to fight. It had descended into Tel Aviv in total darkness:

There hadn't even been a light on the wing. Once again, Amos went into the closet and pulled down his army uniform, now with a captain's insignia on it, and, once again, it fit. At five o'clock the following morning he left.

He had been assigned, with Danny, to the psychology field unit. The unit had grown since the mid-1950s, when Danny had redesigned the selection system. In early 1973 an American psychologist named James Lester, sent by the Office of Naval Research to study Israeli military psychology, wrote a report in which he described the unit they were about to join. Lester marveled at the entire society — a country that had at once the world's strictest driving tests and the world's highest automobile accident rates — but seems to have been struck especially by the faith the Israeli military placed in their psychologists. "Failure rate in the officer course is running at 15–20%," he wrote. "Such confidence does the military have in the mysteries of psychological research that they are asking the Selection Section to try to identify these 15% during the first week in training."

The head of Israeli military psychology, Lester reported, was an oddly powerful character named Benny Shalit. Shalit had argued for, and received, a new, elevated status for military psychology. His unit had a renegade quality to it; Shalit had gone so far as to sew an insignia of his own design onto its uniform. It consisted of the Israeli olive branch and sword, Lester explained, "topped by an eye which symbolizes assessment, insight, or something along those lines." In his attempts to turn his psychology unit into a fighting force, Shalit had dreamed up ideas that struck even the psychologists as wacko. Hypnotizing Arabs and sending them to assassinate Arab leaders, for instance. "He actually did hypnotize one Arab," recalled Daniela Gordon, who served under Shalit in the psychology unit. "They took him to the Jordanian border, and he just ran off."

A rumor among Shalit's subordinates — and it refused to die — was that Shalit kept the personality assessments made of all the Israeli military big shots, back when they were young men entering the army, and let them know that he wouldn't be shy about making them public. Whatever the reason, Benny Shalit had an unusual ability to get his way in the Israeli military. And one of the unusual things Shalit had asked for, and received, was the right to embed psychologists in army units, where they might directly advise commanders. "Field psychologists are in a position to make recommendations on a variety of unconventional issues," Lester reported to his U.S. Navy superiors. "For example, one noticed that infantry troops in hot weather, stopping to open soft drinks with their ammunition magazines, often damaged the stock. It was possible to redesign the stock so that a tool for opening bottles was included." Shalit's psychologists had eliminated the unused sights on submachine guns, and changed the way machine-gun units worked together, to increase the rate at which they fired. Psychologists in the Israeli army were, in short, off the leash. "Military psychology is alive and well in Israel," concluded the United States Navy's reporter on the ground. "It is an interesting question whether or not the psychology of the Israelis is becoming a military one."

What Benny Shalit's field psychologists might do during an actual battle, however, was unclear. "The psychology unit did not have the faintest idea what to do," said Eli Fishoff, who served as Benny Shalit's second-in-command. "The war was totally unexpected. We were just thinking maybe it's the end of us." In a matter of days the Israeli army had lost more men, as a percentage of the population, than the United States military lost in the entire Vietnam War. The war was later described by the Israeli government as a "demographic disaster" because of the prominence and talent of the Israelis who were killed. In the psychology unit someone came up with the idea of designing a questionnaire, to determine what, if anything, might be done to improve the morale of the troops. Amos seized upon it, helped to design the questions, and then used the entire exercise more or less as an excuse to get himself closer to the action. "We just got a jeep and went bouncing around in the Sinai looking for something useful to do," said Danny.

Their fellow psychologists who watched Danny and Amos toss rifles into the back of a jeep and set out for the battlefield thought they were out of their minds. "Amos was so excited — like a little child," recalled Yaffa Singer. "But it was *crazy* for them to go to the Sinai. It was so dangerous. It was absolutely crazy to send them out with those questionnaires." The risk of running directly into enemy tanks and planes was the least of it. There were land mines everywhere; it was easy to get lost. "They didn't have guards," said Daniela Gordon, their commanding officer. "They guarded themselves." All of them felt less concern for Amos than for Danny. "We were very worried about sending Danny on his own," said Eli Fishoff, head of the field psychologists. "I wasn't so worried about Amos — because Amos was a fighter."

The moment Danny and Amos were in the jeep roaring through the Sinai, however, it was Danny who became useful. "He was jumping off the car and grilling people," recalled Fishoff. Amos seemed like the practical one, but Danny, more than Amos, had a gift for finding solutions to problems where others failed even to notice that there was a problem to solve. As they sped toward the front lines, Danny noticed the huge piles of garbage on the roadsides: the leftovers from the canned meals supplied by the U.S. Army. He examined what the soldiers had eaten and what they had thrown out. (They liked the canned grapefruit.) His subsequent recommendation that the Israeli army analyze the garbage and supply the soldiers with what they actually wanted made newspaper headlines.

Israeli tank drivers were just then being killed in action at an unprecedented rate. Danny visited the site where new tank drivers were being trained, as quickly as possible, to replace the ones who had died.

Groups of four men took turns in two-hour shifts on a tank. Danny pointed out that people learn more efficiently in short bursts, and that new tank drivers might be educated faster if the trainees rotated behind the wheel every thirty minutes. He also somehow found his way to the Israeli Air Force. Fighter pilots were also dying in unprecedented numbers, because of Egypt's use of new and improved surface-to-air missiles provided by the Soviet Union. One squadron had suffered especially horrific losses. The general in charge wanted to investigate, and possibly punish, the unit. "I remember him saying accusingly that one of the pilots had been hit 'not only by one missile but by four!' As if that was conclusive evidence of his ineptitude," recalled Danny.

Danny explained to the general that he had a sample size problem: The losses experienced by the supposedly inept fighter squadron could have occurred by random chance alone. If he investigated the unit, he would no doubt find patterns in behavior that might serve as an explanation. Perhaps the pilots in that squadron had paid more visits to their families; or maybe they wore funny-colored underpants. Whatever he found would be a meaningless illusion, however. There weren't enough pilots in the squadron to achieve statistical significance. On top of it, an investigation, implying blame, would be horrible for morale. The only point of an inquiry would be to preserve the general's feelings of omnipotence. The general listened to Danny and stopped the inquiry. "I have considered that my only contribution to the war effort," said Danny.

The actual business at hand — putting questions to soldiers fresh from combat — Danny found pointless. Many of the soldiers were traumatized. "We were wondering what to do with people who were in shock — how even to evaluate them," said Danny. "Every soldier was frightened, but there were some people who couldn't function." Shell-shocked Israeli soldiers resembled people with depression. There were some problems Danny didn't feel equipped to deal with, and this was one of them.

He didn't really want to be in the Sinai anyway, not in the way Amos seemed to want to be there. "I remember a sense of futility — that we were wasting our time there," he said. When their jeep bounced once too often and caused Danny's back to go out, he quit the journey — and left Amos alone to administer the questionnaires. From their jeep rides he retained a single vivid memory. "We went to sleep near a tank," he recalled. "On the ground. And Amos didn't like where I was sleeping, because he thought the tank might move and crush me. And I remember being very, very touched by this. It was not sensible advice. A tank makes a lot of noise. But that he was worried about me."

Later, the Walter Reed Army Institute of Research undertook a study of the war. "Battle Shock Casualties During the 1973 Arab-Israeli War," it was called. The psychiatrists who prepared the report noted that the war was unusual in its intensity — it was fought twenty-four hours a day, at least at the start — and in the losses suffered. The report also noted that, for the first time, Israeli soldiers were diagnosed with psychological trauma. The questionnaires Amos had helped to design asked the soldiers many simple questions: Where were you? What did you do? What did you see? Was the battle a success? If not, why not? "People started to talk about fear," recalls Yaffa Singer. "About their emotions. From the war of independence until 1973 it hadn't been allowed. We are supermen. No one has the guts to talk about fear. If we talk about it maybe we won't survive."

For days after the war, Amos sat with Singer and two other colleagues in the psychology field unit and read through the soldiers' answers to his questions. They spoke of their motives for fighting. "It's such horrible information that people tend to bury it," said Singer. But caught fresh, the soldiers revealed to the psychologists sentiments that, in retrospect, seemed blindingly obvious. "We asked, why is anyone fighting for Israel?" said Singer. "Until that moment we were just patriots. When we started reading the questionnaires it was so obvious: They were fighting for their friends. Or for their families. Not for the nation. Not for Zionism. At the time it was a huge realization." Perhaps for the first time, Israeli soldiers spoke openly of their feelings, as they watched five of their beloved platoon mates blown to bits or as they saw their best friend on earth killed because he turned left when he was supposed to turn right. "It was heartbreaking to read them," said Singer.

Right up until the fighting stopped, Amos sought risks that he didn't need to take — that in fact others thought were foolish to take. "He decided to witness the end of the war along the Suez," recalled Barbara, "even though he knew full well that shelling continued after the time of the cease-fire." Amos's attitude toward physical risk occasionally shocked even his wife. Once, he announced that he wanted to start jumping out of airplanes again, just for fun. "I said you are the father of children," said Barbara, "That ended the discussion." Amos wasn't a thrill seeker, exactly, but he had strong, almost childlike passions that, every so often, he allowed to grab hold of him and take him places most people would never wish to go.

In the end, he crossed the Sinai to the Suez Canal. Rumors circulated that the Israeli army might march all the way to Cairo, and that Soviets were sending nuclear weapons to Egypt to prevent them from doing so. Arriving at the Suez, Amos found that the shelling hadn't merely continued; it had intensified. There was now a longstanding tradition, on both sides of any Arab-Israeli war, of seizing the moment immediately before a formal cease-fire to fire any remaining ammunition at each other. The spirit of the thing was: Kill as many of them as you can, while you can. Wandering around near thc Suez Canal and sensing an incoming missile, Amos leapt into a trench and landed on top of an Israeli soldier.

Are you a bomb? asked the terrified soldier.

No, I'm Amos, said Amos.

So I'm not dead? asked the soldier.

You're not dead, said Amos.

That was the one story Amos told. Apart from that, he seldom mentioned the war again.

In late 1973 or early 1974, Danny gave a talk, which he would deliver more than once, and which he called "Cognitive Limitations and Public Decision Making." It was troubling to consider, he began, "an organism equipped with an affective and hormonal system not much different from that of the jungle rat being given the ability to destroy every living thing by pushing a few buttons." Given the work on human judgment that he and Amos had just finished, he found it further troubling to think that "crucial decisions are made, today as thousands of years ago, in terms of the intuitive guesses and preferences of a few men in positions of authority." The failure of decision makers to grapple with the inner workings of their own minds, and their desire to indulge their gut feelings, made it "quite likely that the fate of entire societies may be sealed by a series of avoidable mistakes committed by their leaders."

Before the war, Danny and Amos had shared the hope that their work on human judgment would find its way into high-stakes real-world decision making. In this new field called decision analysis, they could transform high-stakes decision making into a sort of engineering problem. They would design decision-making *systems.* Experts on decision making would sit with leaders in business, the military, and government and help them to frame every decision explicitly as a gamble; to calculate the odds of this or that happening; and to assign values to every possible outcome. *If we seed the hurricane, there is a 50 percent chance we lower its wind speed but a 5 percent chance that we lull people who really should evacuate into a false sense of security: What do we do?* In the bargain, the decision analysts would remind important decision makers that their gut feelings had mysterious powers to steer them wrong. "The general change in our culture toward numerical formulations will give room for explicit reference to uncertainty," Amos wrote, in notes to himself for a talk of his own. Both Amos and Danny thought that voters and shareholders and all the other people who lived with the consequences of high-level decisions might come to develop a better understanding of the nature of decision making. They would learn to evaluate a decision not by its outcomes — whether it turned out to be right or wrong — but by the process that led to it. The job of the decision maker wasn't to be right but to figure out the odds in any decision and play them well. As Danny told audiences in Israel, what was needed was a "transformation of cultural attitudes to uncertainty and to risk."

Exactly how some decision analyst would persuade any business, military, or political leader to allow him to edit his thinking was unclear. How would you even persuade some important decision maker to assign numbers to his "utilities"? Important people didn't want their gut feelings pinned down, even by themselves. And that was the rub.

Later, Danny recalled the moment he and Amos lost faith in decision analysis. The failure of Israeli intelligence to anticipate the Yom Kippur attack led to an upheaval in the Israeli government and a subsequent brief period of introspection. They'd won the war, but the outcome felt like a loss. The Egyptians, who had suffered even greater losses, were celebrating in the streets as if they had won, while everyone in Israel was trying to figure out what had gone wrong. Before the war, the Israeli intelligence unit had insisted, despite a lot of evidence to the contrary, that Egypt would never attack Israel so long as Israel maintained air superiority. Israel had maintained air superiority, and yet Egypt had attacked. After the war, with the view that perhaps it could do better, Israel's Ministry of Foreign Affairs set up its own intelligence unit. The man in charge of it, Zvi Lanir, sought Danny's help. In the end, Danny and Lanir conducted an elaborate exercise in decision analysis. Its basic idea was to introduce a new rigor in dealing with questions of national security. "We started with the idea that we should get rid of the usual intelligence report," said Danny. "Intelligence reports are in the form of essays. And essays have the characteristic that they can be understood any way you damn well please." In place of the essay, Danny wanted to give Israel's leaders probabilities, in numerical form.

In 1974, U.S. Secretary of State Henry Kissinger had served as the middleman in peace negotiations between Israel and Egypt and between Israel and Syria. As a prod to action, Kissinger had sent the Israeli government the CIA's assessment that, if the attempt to make peace failed, very bad events were likely to follow. Danny and Lanir set out to give Israeli foreign minister Yigal Allon precise numerical estimates of the likelihood of some very specific bad things happening. They assembled a list of possible "critical events or concerns": regime change in Jordan, U.S. recognition of the Palestine Liberation Organization, another full-scale war with Syria, and so on. They then surveyed experts and well-informed observers to establish the likelihood of each event. Among these people, they found a remarkable consensus: There wasn't a lot of disagreement about the odds. When Danny asked the experts what the effect might be of the failure of Kissinger's negotiations on the probability of war with Syria, for instance, their answers clustered around "raises the chance of war by 10 percent."

Danny and Lanir then presented their probabilities to Israel's Foreign Ministry. ("The National Gamble," they called their report.) Foreign Minister Allon looked at the numbers and said, "10 percent increase? — that is a small difference."

Danny was stunned: If a 10 percent increase in the chances of full-scale war with Syria wasn't enough to interest Allon in Kissinger's peace process, how much would it take to turn his head? That number represented the best estimate of the odds. Apparently the foreign minister didn't want to rely on the best estimates. He preferred his own internal probability calculator: his gut. "That was the moment I gave up on decision analysis," said Danny. "No one ever made a decision because of a number. They need a story." As Danny and Lanir wrote, decades later, after the U.S. Central Intelligence Agency asked them to describe their experience in decision analysis, the Israeli Foreign Ministry was "indifferent to the specific probabilities." What was the point of laying out the odds of a gamble, if the person taking it either didn't believe the numbers or didn't want to know them? The trouble, Danny suspected, was that "the understanding of numbers is so weak that they don't communicate anything. Everyone feels that those probabilities are not real — that they are just something on somebody's mind."

In the history of Danny and Amos, there are periods when it is difficult to disentangle their enthusiasm for their ideas from their enthusiasm for each other. The moments before and after the Yom Kippur war appear, in hindsight, less like a natural progression from one idea to the next than two men in love scrambling to find an excuse to be together. They felt they were finished exploring the errors that arose from the rules of thumb people use to evaluate probabilities in any uncertain situation. They'd found decision analysis promising but ultimately futile. They went back and forth on writing a general interest book about the various ways the human mind deals with uncertainty; for some reason, they could never get beyond a sketchy outline and false starts of a few chapters. After the Yom Kippur war — and the ensuing collapse of the public's faith in the judgment of Israeli government officials — they thought that what they really should do was reform the educational system so that future leaders were taught how to think. "We have attempted to teach people to be aware of the pitfalls and fallacies of their own reasoning," they wrote, in a passage for the popular book that never came to be. "We have attempted to teach people at various levels in government, army etc. but achieved only limited success."

Adult minds were too self-deceptive. Children's minds were a different matter. Danny created a course in judgment for elementary school children, Amos briefly taught a similar class to high school students, and they put together a book proposal. "We found these experiences highly encouraging," they wrote. If they could teach Israeli kids how to think — how to detect their own seductive but misleading intuition and to correct for it — who knew where it might lead? Perhaps one day those children would grow up to see the wisdom of encouraging Henry Kissinger's next efforts to make peace between Israel and Syria. But this, too, they never followed through on. They never went broad. It was as if the temptation to address the public interfered with their interest in each other's minds.

Instead, Amos invited Danny to explore the question that had kept Amos interested in psychology: How did people make decisions? "One day, Amos just said, 'We're finished with judgment. Let's do decision making,' " recalled Danny.

The distinction between judgment and decision making appeared as fuzzy as the distinction between judgment and prediction. But to Amos, as to other mathematical psychologists, they were distinct fields of inquiry. A person making a judgment was assigning odds. How likely is it that that guy will be a good NBA player? How risky is that triple-A-rated subprime mortgage–backed CDO? Is the shadow on the X-ray cancer? Not every judgment is followed by a decision, but every decision implies some judgment. The field of decision making explored what people did after they had formed some judgment — after they knew the odds, or thought they knew the odds, or perhaps had judged the odds unknowable. Do I pick that player? Do I buy that CDO? Surgery or chemotherapy? It sought to understand how people acted when faced with risky options.

Students of decision making had more or less given up on real-world investigations and reduced the field to the study of hypothetical gambles, made by subjects in a lab, in which the odds were explicitly stated. Hypothetical gambles played the same role in the study of decision making that the fruit fly played in the study of genetics. They served as proxies for phenomena impossible to isolate in the real world. To introduce Danny to his field — Danny knew nothing about it — Amos gave him an undergraduate textbook on mathematical psychology that he had written with his teacher Clyde Coombs and another Coombs student, Robyn Dawes, the researcher who had confidently and incorrectly guessed "Computer scientist!" when Danny handed him the Tom W. sketch in Oregon. Then he directed Danny to a very long chapter called "Individual Decision Making."

The history of decision theory — the

textbook explained to Danny — began in the early eighteenth century, with dice-rolling French noblemen asking court mathematicians to help them figure out how to gamble. The expected value of a gamble was the sum of its outcomes, each weighted by the probability of its occurring. If someone offers you a coin flip, and you win $100 if the coin lands on heads but lose $50 if it lands on tails, the expected value is $100 × 0.5 + (-$50) × 0.5, or $25. If you follow the rule that you take any bet with a positive expected value, you take the bet. But anyone with eyes could see that people, when they made bets, didn't always act as if they were seeking to maximize their expected value. Gamblers accepted bets with negative expected values; if they didn't, casinos wouldn't exist. And people bought insurance, paying premiums that exceeded their expected losses; if they didn't, insurance companies would have no viable business. Any theory pretending to explain how a rational person should take risks must at least take into account the common human desire to buy insurance, and other cases in which people systematically failed to maximize expected value.

The major theory of decision making, Amos's textbook explained, had been published in the 1730s by a Swiss mathematician named Daniel Bernoulli. Bernoulli sought to account a bit better than simple calculations of expected value for how people actually behaved. "Let us suppose a pauper happens to acquire a lottery ticket by which he may with equal probability win either nothing or 20,000 ducats," he wrote, back when a ducat was a ducat. "Will he have to evaluate the worth of the ticket as 10,000 ducats, or would he be acting foolishly if he sold it for 9,000 ducats?" To explain why a pauper would prefer 9,000 ducats to a 50-50 chance to win 20,000, Bernoulli resorted to sleight of hand. People didn't maximize value, he said; they maximized "utility."

What was a person's "utility"? (That odd, off-putting word here meant something like "the value a person assigns to money.") Well, that depended on how much money the person had to begin with. But a pauper holding a lottery ticket with an expected value of 10,000 ducats would certainly experience greater utility from 9,000 ducats in cash.

"People will choose whatever they most want" is not all that helpful as a theory to predict human behavior. What saved "expected utility theory," as it came to be called, from being so general as to be meaningless were its assumptions about human nature. To his assumption that people making decisions sought to maximize utility, Bernoulli added an assumption that people were "risk averse." Amos's textbook defined risk aversion this way: "The more money one has, the less he values each additional increment, or, equivalently, that the utility of any additional dollar diminishes with an increase in capital." You value the second thousand dollars you get your hands on a bit less than you do the first thousand, just as you value the third thousand a bit less than the second thousand. The marginal value of the dollars you give up to buy fire insurance on your house is less than the marginal value of the dollars you lose if your house burns down — which is why even though the insurance is, strictly speaking, a stupid bet, you buy it. You place less value on the $1,000 you stand to win flipping a coin than you do on the $1,000 already in your bank account that you stand to lose — and so you reject the bet. A pauper places so much value on the first 9,000 ducats he gets his hands on that the risk of not having them overwhelms the temptation to gamble, at favorable odds, for more.

This was not to say that real people in the

real world behaved as they did *because* they had the traits Bernoulli ascribed to them. Only that the theory seemed to describe some of what people did in the real world, with real money. It explained the desire to buy insurance. It distinctly did not explain the human desire to buy a lottery ticket, however. It effectively turned a blind eye to gambling. Odd this, as the search for a theory about how people made risky decisions had started as an attempt to make Frenchmen shrewder gamblers.

Amos's text skipped over the long, tortured history of utility theory after Bernoulli all the way to 1944. A Hungarian Jew named John von Neumann and an Austrian anti-Semite named Oskar Morgenstern, both of whom fled Europe for America, somehow came together that year to publish what might be called the rules of rationality. A rational person making a decision between risky propositions, for instance, shouldn't violate the von Neumann and Morgenstern transitivity axiom: If he preferred *A* to *B* and *B* to *C*, then he should prefer *A* to *C*. Anyone who preferred *A* to *B* and *B* to *C* but then turned around and preferred *C* to *A* violated expected utility theory. Among the remaining rules, maybe the most critical — given what would come

— was what von Neumann and Morgenstern called the "independence axiom." This rule said that a choice between two gambles shouldn't be changed by the introduction of some irrelevant alternative. For example: You walk into a deli to get a sandwich and the man behind the counter says he has only roast beef and turkey. You choose turkey. As he makes your sandwich he looks up and says, "Oh, yeah, I forgot I have ham." And you say, "Oh, then I'll take the roast beef." Von Neumann and Morgenstern's axiom said, in effect, that you can't be considered rational if you switch from turkey to roast beef just because they found some ham in the back.

And, really, who would switch? Like the other rules of rationality, the independence axiom seemed reasonable, and not obviously contradicted by the way human beings generally behaved.

Expected utility theory was just a theory. It didn't pretend to be able to explain or predict everything people did when they faced some risky decision. Danny gleaned its importance not from reading Amos's description of it in the undergraduate textbook but only from the way Amos spoke of it. "This was a sacred thing for Amos," said Danny. Although the theory made no great claim to psychological truth, the textbook Amos had coauthored made it clear that it had been accepted as psychologically true. Pretty much everyone interested in such things, a group that included the entire economics profession, seemed to take it as a fair description of how ordinary people faced with risky alternatives actually went about making choices. That leap of faith had at least one obvious implication for the sort of advice economists gave to political leaders: It tilted everything in the direction of giving people the freedom to choose and leaving markets alone. After all, if people could be counted on to be basically rational, markets could, too.

Amos had clearly wondered about that, even as a Michigan graduate student. Amos had always had an almost jungle instinct for the vulnerability of other people's ideas. He of course knew that people made decisions that the theory would not have predicted. Amos himself had explored how people could be — as the theory assumed they were not — reliably "intransitive." As a graduate student in Michigan, he had induced both Harvard undergraduates and convicted murderers in Michigan prisons, over and over again, to choose gamble *A* over gamble *B,* then choose gamble *B* over gamble *C* —

and then turn around and choose *C* instead of *A*. That violated a rule of expected utility theory. And yet Amos had never followed his doubts very far. He saw that people sometimes made mistakes; he did not see anything systematically irrational in the way they made decisions. He hadn't figured out how to bring deep insights about human nature into the mathematical study of human decision making.

By the summer of 1973, Amos was searching for ways to undo the reigning theory of decision making, just as he and Danny had undone the idea that human judgment followed the precepts of statistical theory. On a trip to Europe with his friend Paul Slovic, he shared his latest thoughts about how to make room, in the world of decision theory, for a messier view of human nature. "Amos warns against pitting utility theory vs. an alternative model in a direct, head to head, empirical test," Slovic relayed, in a letter to a colleague, in September 1973. "The problem is that utility theory is so general that it is hard to refute. Our strategy should be to take the offensive in building a case, not against utility theory, but *for* an alternative conception that brings man's limitations in as a constraint."

Amos had at his disposal a connoisseur of

man's limitations. He now described Danny as "the world's greatest living psychologist." Not that he ever said anything so flattering to Danny directly. ("Manly reticence was the rule," said Danny.) He never fully explained to Danny why he thought to invite him into decision theory — a technical and antiseptic field Danny cared little about and knew less of. But it is hard to believe that Amos was simply looking around for something else they might do together. It's easier to believe that Amos suspected what might happen after he gave Danny his textbook on the subject. That moment has the feel of an old episode of *The Three Stooges,* when Larry plays "Pop Goes the Weasel" and triggers Curly into a frenzy of destruction.

Danny read Amos's textbook the way he might have read a recipe written in Martian. He decoded it. He had long ago realized that he wasn't a natural applied mathematician, but he could follow the logic of the equations. He knew that he was meant to respect, even revere, them. Amos was a member of high standing in the society of mathematical psychologists. That society in turn looked down upon much of the rest of psychology. "It is a given that people who use mathematics have some glamour," said

Danny. "It was prestigious because it borrowed the aura of mathematics and because nobody else could understand what was going on there." Danny couldn't escape the growing prestige of math in the social sciences: His remove counted against him. But he didn't really admire decision theory, or care about it. He cared why people behaved as they did. And to Danny's way of thinking, the major theory of decision making did not begin to describe how people made decisions.

It must have come as something of a relief to him, as he neared the end of Amos's chapter on expected utility theory, to arrive at the following sentence: "Some people, however, remained unconvinced by the axioms."

One such person, the textbook went on to say, was Maurice Allais. Allais was a French economist who disliked the self-certainty of American economists. He especially disapproved of the growing tendency in economics, after von Neumann and Morgenstern built their theory, to treat a math model of human behavior as an accurate description of how people made choices. At a convention of economists in 1953, Allais offered what he imagined to be a killer argument against expected utility theory. He asked his audience to imagine their choices in the following two situations (the dollar amounts used by Allais are here multiplied by ten to account for inflation and capture the feel of his original problem):

Situation 1. You must choose between having:

1) $5 million for sure

or this gamble

2) An 89 percent chance of winning $5 million
A 10 percent chance of winning $25 million
A 1 percent chance to win zero

Most people who looked at that, apparently including many of the American economists in Allais's audience, said, "Obviously, I'll take door number 1, the $5 million for sure." They preferred the certainty of being rich to the slim possibility of being even richer. To which Allais replied, "Okay, now consider this second situation."

Situation 2. You must choose between having:

3) An 11 percent chance of winning $5 million, with an 89 percent chance to win zero

or

4) A 10 percent chance of winning $25 million, with a 90 percent chance to win zero

Most everyone, including American economists, looked at this choice and said, "I'll take number 4." They preferred the slightly lower chance of winning a lot more money. There was nothing wrong with this; on the face of it, both choices felt perfectly sensible. The trouble, as Amos's textbook explained, was that "this seemingly innocent pair of preferences is incompatible with utility theory." What was now called the Allais paradox had become the most famous contradiction of expected utility theory. Allais's problem caused even the most cold-blooded American economist to violate the rules of rationality.*

Amos's introduction to mathematical

* I apologize for this, but it must be done. Those whose minds freeze when confronted with algebra can skip what follows. A simpler proof of the paradox, devised by Danny and Amos, will come

psychology sketched the controversy and argument that had ensued after Allais posed his paradox. On the American end, the argument was spearheaded by a brilliant American statistician and mathematician named L. J. (Jimmie) Savage, who had made important contributions to utility theory and who admitted that he, too, had been suckered by Allais into contradicting himself. Savage found an even more complicated way to restate Allais's gambles so that at least a few devotees of expected utility

---

later. But here, more or less reproduced from *Mathematical Psychology: An Elementary Introduction,* is the proof of Allais's point that Amos asked Danny to ponder.

Let u stand for utility.

In situation 1: $u(\text{gamble 1}) > u(\text{gamble 2})$

and hence $1u(5) > .10u(25) + .89u(5) + .01u(0)$

so $.11u(5) > .10u(25) + .01u(0)$

Now turn to situation 2, where most people chose 4 over 3. This implies $u(\text{gamble 4}) > u(\text{gamble 3})$

and hence $.10u(25) + .90u(0) > .11u(5) + .89u(0)$

so $.10u(25) + .01u(0) > .11u(5)$

Or the exact reverse of the choice made in the first gamble.

theory, himself included, looked at the second situation and picked option number 3 instead of option number 4. That is, he demonstrated — or thought he had demonstrated — that the Allais "paradox" was not a paradox at all, and that people behaved just as expected utility predicted they would behave. Amos, along with pretty much everyone else who took an interest in such things, remained dubious.

As Danny read up on decision theory, Amos helped him to understand what was important about it and what was not. "He just had impeccable taste," said Danny. "He knew what the problems were. He knew how to situate himself in the broad field. I didn't have that." What was important, Amos said, were the unresolved puzzles. "Amos said, 'This is the story, this is the game. The game is to solve the Allais paradox.' "

Danny wasn't inclined to see the paradox as a problem of logic. It looked to him more like a quirk in human behavior. "I wanted to understand the psychology of what was going on," he said. He sensed that Allais himself hadn't given much thought to why people might choose in a way that violated the major theory of decision making. But to Danny the reason seemed obvious: regret.

In the first situation people sensed that they would look back on their decision, if it turned out badly, and feel they had screwed up; in the second situation, not so much. Anyone who turned down a certain gift of $5 million would experience far more regret, if he wound up with nothing, than a person who turned down a gamble in which he stood a slight chance of winning $5 million. If people mostly chose option 1, it was because they sensed the special pain they would experience if they chose option 2 and won nothing. Avoiding that pain became a line item on the inner calculation of their expected utility. Regret was the ham in the back of the deli that caused people to switch from turkey to roast beef.

Decision theory had approached the seeming contradiction at the heart of the Allais paradox as a technical problem. Danny found that silly: There was no contradiction. There was just psychology. The understanding of any decision had to account not just for the financial consequences but for the emotional ones, too. "Obviously it is not regret itself that determines decisions — no more than the actual emotional response to consequences ever determines the prior choice of a course of action," Danny wrote to Amos, in one of a series of memos on the subject. “It is the anticipation of regret that affects decisions, along with the anticipation of other consequences.” Danny thought that people anticipated regret, and adjusted for it, in a way they did not anticipatc or adjust for other emotions. “What might have been is an essential component of misery,’ ” he wrote to Amos. “There is an asymmetry here, because considerations of how much worse things could have been is not a salient factor in human joy and happiness.”

Happy people did not dwell on some imagined unhappiness the way unhappy people imagined what they might have done differently so that they might be happy. People did not seek to avoid other emotions with the same energy they sought to avoid regret.

When they made decisions, people did not seek to maximize utility. They sought to minimize regret. As the starting point for a new theory, it sounded promising. When people asked Amos how he made thc big decisions in his life, he often told them that his strategy was to imagine what he would come to regret, after he had chosen some option, and to choose the option that would make him feel the least regret. Danny, for his part, personified regret. Danny would resist a change to his airline reservations, even when the change made his life a lot easier, because he imagined the regret he would feel if the change led to some disaster. It's not a stretch to say that Danny anticipated anticipating regret. He was perfectly capable of anticipating the regret provoked by events that might never occur and decisions that he might never need to make. Once, at a dinner with Amos and their wives, Danny went on at length and with great certainty about his premonition that his son, then still a boy, would one day join the Israeli military; that war would break out; and that his son would be killed. "What were the odds of all that happening?" said Barbara Tversky. "Minuscule. But I couldn't talk him out of it. It was so unpleasant talking with him about these small probabilities that I just gave up." It was as if Danny thought that by anticipating his feelings he might dull the pain they would inevitably bring.

By the end of 1973, Amos and Danny were spending six hours a day with each other, either holed up in a conference room or on long walks across Jerusalem. Amos hated smoke; he hated being around people who smoked. Danny was still smoking two packs of cigarettes a day, and yet Amos never said a word. All that mattered was the conversation. When they weren't with each other, they were writing memos to each other, to clarify and extend what had been said. If they happened to find themselves at the same social function, they inevitably wound up in the corner of a room, talking to each other. "We just found each other more interesting than anyone else," said Danny. "Even if we had just spent the entire day working together." They'd become a single mind, creating ideas about why people did what they did, and cooking up odd experiments to test them. For instance, they put this scenario to subjects:

> You have participated in a lottery at a fair, and have bought a single expensive ticket in the hope of winning the single large prize that is offered. The ticket was drawn blindly from a large urn, and its number is 107358. The results of the lottery are now announced, and it turns out that the winning number is 107359.

They asked their subjects to rate their unhappiness on a scale from 1 to 20. Then they went to two other groups of subjects and gave them the same scenario, but with one change: the winning number. One group of subjects was told that the winning number was 207358; the second group was told that the winning number was 618379. The first group professed greater unhappiness than the second. Weirdly — but as Danny and Amos had suspected — the further the winning number was from the number on a person's lottery ticket, the less regret they felt. "In defiance of logic, there is a definite sense that one comes closer to winning the lottery when one's ticket number is similar to the number that won," Danny wrote in a memo to Amos, summarizing their data. In another memo, he added that "the general point is that the same state of affairs (objectively) can be experienced with very different degrees of misery," depending on how easy it is to imagine that things might have turned out differently.

Regret was sufficiently imaginable that people conjured it out of situations they had no control over. But it was of course at its most potent when people might have done something to avoid it. What people regretted, and the intensity with which they regretted it, was not obvious.

War and politics were never far from Amos and Danny's minds or their conversations. They watched their fellow Israelis closely in the aftermath of the Yom Kippur war. Most regretted that Israel had been caught by surprise. Some regretted that Israel had not attacked first. Few regretted what both Danny and Amos thought they should most regret: the Israeli government's reluctance to give back the territorial gains from the 1967 war. Had Israel given back the Sinai to Egypt, Sadat would quite likely never have felt the need to attack in the first place. Why didn't people regret Israel's inaction? Amos and Danny had a thought: People regretted what they had done, and what they wished they hadn't done, far more than what they had not done and perhaps should have. "The pain that is experienced when the loss is caused by an act that modified the status quo is significantly greater than the pain that is experienced when the decision led to the retention of the status quo," Danny wrote in a memo to Amos. "When one fails to take action that could have avoided a disaster, one does not accept responsibility for the occurrence of the disaster."

They set out to build a theory of regret. They were uncovering, or thought they were uncovering, what amounted to the rules of regret. One rule was that the emotion was closely linked to the feeling of "coming close" and failing. The nearer you came to achieving a thing, the greater the regret you experienced if you failed to achieve it.* A second rule: Regret was closely linked to feelings of responsibility. The more control you felt you had over the outcome of a gamble, the greater the regret you experienced if the gamble turned out badly. People anticipated regret in Allais's problem not from the failure to win a gamble but from the decision to forgo a certain pile of money.

That was another rule of regret. It skewed any decision in which a person faced a choice between a sure thing and a gamble. This tendency was not merely of academic interest. Danny and Amos agreed that there was a real-world equivalent of a "sure

* Two decades later, in 1995, the American psychologist Thomas Gilovich, who collaborated in turn with Danny and Amos, coauthored a study that examined the relative happiness of silver and bronze medal winners at the 1992 Summer Olympics. From video footage, subjects judged the bronze medal winners to be happier than the silver medal winners. The silver medalists, the authors suggested, dealt with the regret of not having won gold, while the bronze medalists were just happy to be on a podium.

thing": the status quo. The status quo was what people assumed they would get if they failed to take action. "Many instances of prolonged hesitation, and of continued reluctance to take positive action, should probably be explained in this fashion," wrote Danny to Amos. They played around with the idea that the anticipation of regret might play an even greater role in human affairs than it did if people could somehow know what would have happened if they had chosen differently. "The absence of definite information concerning the outcomes of actions one has not taken is probably the single most important factor that keeps regret in life within tolerable bounds," Danny wrote. "We can never be absolutely sure that we would have been happier had we chosen another profession or another spouse. . . . Thus, we are often protected from painful knowledge concerning the quality of our decisions."

They spent more than a year working and reworking the same basic idea: In order to explain the paradoxes that expected utility could not explain, and create a better theory to predict behavior, you had to inject psychology into the theory. By testing how people choose between various sure gains and gains that were merely probable, they traced the contours of regret.

> Which of the following two gifts do you prefer?
> Gift A: A lottery ticket that offers a 50 percent chance of winning $1,000
> Gift B: A certain $400
>
> or
>
> Which of the following gifts do you prefer?
> Gift A: A lottery ticket that offers a 50 percent chance of winning $1 million
> Gift B: A certain $400,000

They collected great heaps of data: choices people had actually made. "Always keep one hand firmly on data," Amos liked to say. Data was what set psychology apart from philosophy, and physics from metaphysics. In the data, they saw that people's subjective feelings about money had a lot in common with their perceptual experiences. People in total darkness were extremely sensitive to the first glimmer of light, just as people in total silence were alive to the faintest sound, and people in tall buildings were quick to detect even the slightest swaying. As you turned up the lights or the sound or the movement, people became less sensitive to incremental change. So, too, with money. People felt greater pleasure going from 0 to $1 million than they felt going from $1 million to $2 million. Of course, expected utility theory also predicted that people would take a sure gain over a bet that offered an expected value of an even bigger gain. They were "risk averse." But what was this thing that everyone had been calling "risk aversion?" It amounted to a fee that people paid, willingly, to avoid regret: a regret premium.

Expected utility theory wasn't exactly wrong. It simply did not understand itself, to the point where it could not defend itself against seeming contradictions. The theory's failure to explain people's decisions, Danny and Amos wrote, "merely demonstrates what should perhaps be obvious, that nonmonetary consequences of decisions cannot be neglected, as they all too often are, in applications of utility theory." Still, it wasn't obvious how to weave what amounted to a collection of insights about an emotion into a theory of how people make risky decisions. They were groping. Amos liked to use an expression he'd read someplace: "carving nature at its joint." They were trying to carve human nature at its joint, but the joints of an emotion were elusive. That was one reason Amos didn't particularly like to think or talk about emotion; he didn't like things that were hard to measure. "This is indeed a complex theory," Danny confessed one day in a memo. "In fact it consists of several mini-theories, which are rather loosely connected."

In reading about expected utility theory, Danny had found the paradox that purported to contradict it not terribly puzzling. What puzzled Danny was what the theory had left out. "The smartest people in the world are measuring utility," he recalled. "As I'm reading about it, something strikes me as really, really peculiar." The theorists seemed to take it to mean "the utility of having money." In their minds, it was linked to *levels* of wealth. More, because it was more, was always better. Less, because it was less, was always worse. This struck Danny as false. He created many scenarios to show just how false it was:

> Today Jack and Jill each have a wealth of 5 million.
>
> Yesterday, Jack had 1 million and Jill had 9 million.
>
> Are they equally happy? (Do they have

the same utility?)

Of course they weren't equally happy. Jill was distraught and Jack was elated. Even if you took a million away from Jack and left him with less than Jill, he'd still be happier than she was. In people's perceptions of money, as surely as in their perception of light and sound and the weather and everything else under the sun, what mattered was not the absolute levels but *changes.* People making choices, especially choices between gambles for small sums of money, made them in terms of gains and losses; they weren't thinking about absolute levels. "I came back to Amos with that question, expecting that he would explain it to me," Danny recalled. "Instead Amos says, 'You're right.' "

10
The Isolation Effect

It was seldom possible for Amos and Danny to recall where their ideas had come from. They both found it pointless to allocate credit, as their thoughts felt like some alchemical by-product of their interaction. Yet, on occasion, their origins were preserved. The notion that people making risky decisions were especially sensitive to change pretty clearly had at least started with Danny. But it became seriously valuable only because of what Amos said next. One day, toward the end of 1974, as they looked over the gambles they had put to their subjects, Amos asked, "What if we flipped the signs?" Till that point, the gambles had all involved choices between gains. *Would you rather have $500 for sure or a 50-50 shot at $1,000?* Now Amos asked, "What about losses?" As in:

Which of the following do you prefer?

Gift A: A lottery ticket that offers a 50 percent chance of losing $1,000
Gift B: A certain loss of $500

It was instantly obvious to them that if you stuck minus signs in front of all these hypothetical gambles and asked people to reconsider them, they behaved very differently than they had when faced with nothing but possible gains. "It was a eureka moment," said Danny. "We immediately felt like fools for not thinking of that question earlier." When you gave a person a choice between a gift of $500 and a 50-50 shot at winning $1,000, he picked the sure thing. Give that same person a choice between *losing* $500 for sure and a 50-50 risk of losing $1,000, and he took the bet. He became a risk seeker. The odds that people demanded to accept a certain loss over the chance of some greater loss crudely mirrored the odds they demanded to forgo a certain gain for the chance of a greater gain. For example, to get people to prefer a 50-50 chance of $1,000 over some certain gain, you had to lower the certain gain to around $370. To get them to prefer a certain loss to a 50-50 chance of losing $1,000, you had to lower the loss to around $370.

Actually, they soon discovered, you had to

reduce the amount of the certain loss even further if you wanted to get people to accept it. When choosing between sure things and gambles, people's desire to avoid loss exceeded their desire to secure gain.

The desire to avoid loss ran deep, and expressed itself most clearly when the gamble came with the possibility of both loss and gain. That is, when it was like most gambles in life. To get most people to flip a coin for a hundred bucks, you had to offer them far better than even odds. If they were going to lose $100 if the coin landed on heads, they would need to win $200 if it landed on tails. To get them to flip a coin for ten thousand bucks, you had to offer them even better odds than you offered them for flipping it for a hundred. "The greater sensitivity to negative rather than positive changes is not specific to monetary outcomes," wrote Amos and Danny. "It reflects a general property of the human organism as a pleasure machine. For most people, the happiness involved in receiving a desirable object is smaller than the unhappiness involved in losing the same object."

It wasn't hard to imagine why this might be — a heightened sensitivity to pain was helpful to survival. "Happy species endowed with infinite appreciation of pleasures and low sensitivity to pain would probably not survive the evolutionary battle," they wrote.

As they sorted through the implications of their new discovery, one thing was instantly clear: Regret had to go, at least as a theory. It might explain why people made seemingly irrational decisions to accept a sure thing over a gamble with a far greater expected value. It could not explain why people facing losses became risk seeking. Anyone who wanted to argue that regret explains why people prefer a certain $500 to an equal chance to get $0 and $1,000 would never be able to explain why, if you simply subtracted $1,000 from all the numbers and turned the sure thing into a $500 loss, people would prefer the gamble. Amazingly, Danny and Amos did not so much as pause to mourn the loss of a theory they'd spent more than a year working on. The speed with which they simply walked away from their ideas about regret — many of them obviously true and valuable — was incredible. One day they are creating the rules of regret as if those rules might explain much of how people made risky decisions; the next, they have moved on to explore a more promising theory, and don't give regret a second thought.

Instead they set out to determine precisely

where and how people responded to the odds of various bets involving both losses and gains. Amos liked to call good ideas "raisins." There were three raisins in the new theory. The first was the realization that people responded to changes rather than absolute levels. The second was the discovery that people approached risk very differently when it involved losses than when it involved gains. Exploring people's responses to specific gambles, they found a third raisin: People did not respond to probability in a straightforward manner. Amos and Danny already knew, from their thinking about regret, that in gambles that offered a certain outcome, people would pay dearly for that certainty. Now they saw that people reacted differently to different degrees of uncertainty. When you gave them one bet with a 90 percent chance of working out and another with a 10 percent chance of working out, they did not behave as if the first was nine times as likely to work out as the second. They made some internal adjustment, and acted as if a 90 percent chance was actually slightly less than a 90 percent chance, and a 10 percent chance was slightly more than a 10 percent chance. They responded to probabilities not just with reason but with emotion.

Whatever that emotion was, it became stronger as the odds became more remote. If you told them that there was a one-in-a-billion chance that they'd win or lose a bunch of money, they behaved as if the odds were not one in a billion but one in ten thousand. They feared a one-in-a-billion chance of loss more than they should and attached more hope to a one-in-a-billion chance of gain than they should. People's emotional response to extremely long odds led them to reverse their usual taste for risk, and to become risk seeking when pursuing a long-shot gain and risk avoiding when faced with the extremely remote possibility of loss. (Which is why they bought both lottery tickets and insurance.) "If you think about the possibilities at all, you think of them too much," said Danny. "When your daughter is late and you worry, it fills your mind even when you know there is very little to fear." You'd pay more than you should to rid yourself of that worry.

People treated all remote probabilities as if they were possibilities. To create a theory that would predict what people actually did when faced with uncertainty, you had to "weight" the probabilities, in the way that people did, with emotion. Once you did that, you could explain not only why people bought insurance and lottery tickets. You could even explain the Allais paradox.*

At some point, Danny and Amos became

* Here is the simpler version of the paradox. Danny and Amos created it to show how the apparent contradiction might be resolved using their findings about people's attitudes toward probabilities. And so in a funny way they "solved" the Allais paradox twice — once by explaining it with regret, this time by explaining it with their new theory:

You are offered a choice between:

1. $30,000 for sure
2. A gamble that has a 50 percent chance of winning $70,000 and a 50 percent chance of winning nothing

Most people took the $30,000. That was interesting in itself. It showed what was meant by "risk aversion." People choosing between a bet and a certain amount would accept a certain amount that was less than the expected value of the bet (which here is $35,000). That did not violate utility theory. It just meant that the utility of a chance to win 70 grand is less than the utility of a twice as likely chance to win 30 grand — which in this case makes the 30 grand a certainty. But now consider a second choice between bets:

aware that they had a problem on their hands. Their theory explained all sorts of things that expected utility failed to explain. But it implied, as utility theory never had, that it was as easy to get people to take risks as it was to get them to avoid them. All you had to do was present them with a choice that involved a loss. In the more than two

---

1. A gamble that gives you a 4 percent chance to win $30,000 and a 96 percent chance to win nothing
2. A gamble that gives you a 2 percent chance to win $70,000 and a 98 percent chance to win nothing

Most people here preferred 2, the lower chance to win more. But that implied that the "utility" of a chance to win $70,000 is greater than the utility of a twice as likely chance to win $30,000 — or the opposite of the preferences in the first choice. In Danny and Amos's working theory, the paradox was now resolved differently. It wasn't that (or at least not only that) people anticipated regret when making a decision in the first situation that they did not anticipate in making the second. It was that they treated 50 percent as more than 50 percent and saw the difference between 4 percent and 2 percent as far less than it was.

hundred years since Bernoulli started the discussion, intellectuals had regarded risk-seeking behavior as a curiosity. If risk seeking was woven into human nature, as Danny and Amos's theory implied that it was, why hadn't people noticed it before?

The answer, Amos and Danny now thought, was that intellectuals who studied human decision making had been looking in the wrong places. Mostly they had been economists, who directed their attention to the way people made decisions about money. "It is an ecological fact," wrote Amos and Danny in a draft, "that most decisions in that context (except insurance) involve mainly favorable prospects." The gambles that economists studied were, like most savings and investment decisions, choices between gains. In the domain of gains, people were indeed risk averse. They took the sure thing over the gamble. Danny and Amos thought that if the theorists had spent less time with money and more time with politics and war, or even marriage, they might have come to different conclusions about human nature. In politics and war, as in fraught human relationships, the choice faced by the decision maker was often between two unpleasant options. "A very different view of man as a decision maker might well have emerged if the outcomes of decisions in the private-personal, political or strategic domains had been as easily measurable as monetary gains and losses," they wrote.

Danny and Amos spent the first half of 1975 getting their theory into shape so that a rough draft might be shown to other people. They started with the working title "Value Theory" but then changed it to "Risk-Value Theory." For a pair of psychologists who were attacking a theory erected and defended mainly by economists, they wrote with astonishing aggression and confidence. The old theory, they wrote, didn't really even consider how actual human beings grappled with risky decisions. All it did was "to explain risky choices solely in terms of attitudes to money or wealth." Between the lines, the reader could detect their giddiness. "Amos and I are in the middle of our most productive period ever," Danny wrote to Paul Slovic, in early 1975. "We're developing what appears to us to be a rather complete and quite novel account of choice under uncertainty. The regret treatment has been superseded by a sort of reference level or adaptation level treatment." Six months later, Danny wrote Slovic that they had a prototype of a new theory of decision making. "Amos and I barely managed to finish a paper on risky choice in time to present it to an illustrious group of economists who convene in Jerusalem this week," he wrote. "It is still fairly rough."

The meeting in question, billed as a conference on public economics, convened in June 1975 at a kibbutz just outside Jerusalem. And so it was on a farm that a theory that would become among the most influential in the history of economics made its public debut. Decision theory was Amos's field, and so Amos did all the talking. The audience contained at least three current and future Nobel Prize winners in economics: Peter Diamond, Daniel McFadden, and Kenneth Arrow. "When you listened to Amos, you knew you were talking to a first-rate mind," said Arrow. "You raise a question. He's thought of the question already, and he has an answer."

After he listened to Amos's presentation, Arrow had one big question for Amos: What is a loss?

The theory obviously turned on the stark difference in people's feelings when they faced potential losses rather than potential gains. A loss, according to the theory, was when a person wound up worse off than his "reference point." But what was this reference point? The easy answer was: wherever you started from. Your status quo. A loss was just when you ended up worse than your status quo. But how did you determine any person's status quo? "In the experiments it's pretty clear what a loss is," Arrow said later. "In the real world it's not so clear."

Wall Street trading desks at the end of each year offer a flavor of the problem. If a Wall Street trader expects to be paid a bonus of one million dollars and he's given only half a million, he feels himself to be, and behaves as if he is, in the domain of losses. His reference point is an *expectation* of what he would receive. That expectation isn't a stable number; it can be changed in all sorts of ways. A trader who expects to be given a million-dollar bonus, and who further expects everyone else on his trading desk to be given million-dollar bonuses, will not maintain the same reference point if he learns that everyone else just received two million dollars. If he is then paid a million dollars, he is back in the domain of losses. Danny would later use the same point to explain the behavior of apes in experiments researchers had conducted on bonobos. "If both my neighbor in the next cage and I get a cucumber for doing a great job, that's great. But if he gets a banana and I get a cucumber, I will throw the cucumber at the experimenter's face." The moment one ape got a banana, it became the ape next door's reference point.

The reference point was a state of mind. Even in straight gambles you could shift a person's reference point and make a loss seem like a gain, and vice versa. In so doing, you could manipulate the choices people made, simply by the way they were described. They gave the economists a demonstration of the point:

> Problem A. In addition to whatever you own, you have been given $1,000. You are now required to choose between the following options:
>
> Option 1. A 50 percent chance to win $1000
> Option 2. A gift of $500

Most everyone picked option 2, the sure thing.

> Problem B. In addition to whatever you own, you have been given $2,000. You are now required to choose between the fol-

lowing options:

Option 3. A 50 percent chance to lose $1,000
Option 4. A sure loss of $500

Most everyone picked option 3, the gamble.

The two questions were effectively identical. In both cases, if you picked the gamble, you wound up with a 50-50 shot at being worth $2,000. In both cases, if you picked the sure thing, you wound up being worth $1,500. But when you framed the sure thing as a loss, people chose the gamble. When you framed it as a gain, people picked the sure thing. The reference point — the point that enabled you to distinguish between a gain and a loss — wasn't some fixed number. It was a psychological state. "What constitutes a gain or a loss depends on the representation of the problem and on the context in which it arises," the first draft of "Value Theory" rather loosely explained. "We propose that the present theory applies to the gains and losses as *perceived by the subject.*"

Danny and Amos were trying to show that people faced with a risky choice failed to put it in context. They evaluated it in isolation. In exploring what they now called the isolation effect, Amos and Danny had stumbled upon another idea — and its real-world implications were difficult to ignore. This one they called "framing." Simply by changing the description of a situation, and making a gain seem like a loss, you could cause people to completely flip their attitude toward risk, and turn them from risk avoiding to risk seeking. "We invented framing without realizing we were inventing framing," said Danny. "You take two things that should be identical — the way they differ should be irrelevant — and by showing it isn't irrelevant, you show that expected utility theory is wrong." Framing, to Danny, felt like their work on judgment. *Here, look, yet another strange trick the mind played on itself.*

Framing was just another phenomenon: There was never going to be a theory of framing. But Amos and Danny would eventually spend all kinds of time and energy dreaming up examples of the phenomenon, to illustrate how it might distort real-world decisions. The most famous was the Asian Disease Problem.

The Asian Disease Problem was actually two problems, which they gave, separately, to two different groups of subjects innocent of the power of framing. The first group got this problem:

> Problem 1. Imagine that the U.S. is preparing for the outbreak of an unusual Asian disease, which is expected to kill 600 people. Two alternative programs to combat the disease have been proposed. Assume that the exact scientific estimate of the consequence of the programs is as follows:
> If Program A is adopted, 200 people will be saved.
> If Program B is adopted, there is a 1/3 probability that 600 people will be saved, and a 2/3 probability that no people will be saved.
> Which of the two programs would you favor?

An overwhelming majority chose Program A, and saved 200 lives with certainty.

The second group got the same setup but with a choice between two other programs:

> If Program C is adopted, 400 people will die.
> If Program D is adopted, there is a 1/3 probability that nobody will die and a 2/3 probability that 600 people will die.

When the choice was framed this way, an overwhelmingly majority chose Program D. The two problems were identical, but, in the first case, when the choice was framed as a gain, the subjects elected to save 200 people for sure (which meant that 400 people would die for sure, though the subjects weren't thinking of it that way). In the second case, with the choice framed as a loss, they did the reverse, and ran the risk that they'd kill everyone.

People did not choose between things. They chose between descriptions of things. Economists, and anyone else who wanted to believe that human beings were rational, could rationalize, or try to rationalize, loss aversion. But how did you rationalize this? Economists assumed that you could simply measure what people wanted from what they chose. But what if what you want changes with the context in which the options are offered to you? "It was a funny point to make because the point within psychology would have been banal," the psychologist Richard Nisbett later said. "Of course we are affected by how the decision is presented!"

After the meeting between the American economists and the Israeli psychologists on the Jerusalem kibbutz, the economists returned to the United States and Amos sent a letter to Paul Slovic. "Everything considered we got a very favorable response," he wrote. "Somehow, the economists felt that we are right and at the same time they wished we weren't because the replacement of utility theory by the model we outlined would cause them no end of problems."

There was at least one economist who didn't feel that way, but he wasn't, at least when he came upon Danny and Amos's theory, anyone's idea of a future Nobel Prize winner. His name was Richard Thaler. In 1975, Thaler was a thirty-year-old assistant professor in the School of Management at the University of Rochester with vague prospects. It was a wonder he was even there. He had two deeply pronounced traits that rendered him unsuited not just to economics but to academic life. The first was that he was easily bored, and highly imaginative in his attempts to escape boredom. As a child he routinely changed the rules of the games he was expected to play. The first hour and a half of Monopoly, when players march around the board randomly landing on properties and buying them, he found tedious. After playing a few times, he announced, "This is a stupid game." He said that he would only play if all the properties were shuffled and dealt to the players at the start of the game. Same with Scrabble. Finding it boring when he got dealt five "E"s and no high-value consonants, he changed the rules so that the letters were organized into three buckets: vowels, common consonants, and rare, high-value consonants. Each player got the same number of each; after seven rounds, each player was given a high-value consonant. All the changes Thaler made to the games he played as a kid reduced the waiting-around time, and the role of luck, and increased the challenge and, usually, the players' competitiveness.

This was odd, as Thaler's other pronounced trait was a sense of ineptitude. When he was ten or eleven years old, and a B student, his father, a detail-oriented insurance executive, had grown so frustrated with his sloppy schoolwork that he handed his son *The Adventures of Tom Sawyer* and told him to copy a few pages exactly as Mark Twain had written them. Thaler tried, seriously. "I did it over and over, kicking and screaming." Each time, his father found errors — missing words, missing commas. The quotation marks in an exchange between

Tom and Aunt Polly confounded him. Looking back on it, he could see that his problem was more than a lack of effort: He was probably mildly dyslexic. But people just assumed he was either careless or lazy, or both.

And so he began to think of himself this way, too. Economics just then wasn't perhaps the ideal place for people who were easily bored and had trouble with details. Thaler had gone from college straight to graduate school mainly because his father's life had persuaded him that business careers were mind-crushingly boring, and that he had no ability to work for someone else. He couldn't think of what else to do but go to graduate school, and he picked economics because "it seemed kind of practical." Only then did he discover that the field placed a terrifying premium on both precision and mathematical ability — to the point where it seemed that the only people who were allowed to make jokes in their journal articles were the guys who were best at math. By the time Thaler arrived at the University of Rochester's Graduate School of Management, he sat at some distance from his own field, and from his fellow graduate students. "I was more interesting than them, and not as good at math," he said. "What was I good at? It was at finding things that were interesting."

He wrote his thesis about why the infant mortality rate in the United States was twice as high for black as for white populations. Controlling for all the obvious variables — education and income of the parents, whether the baby was born in a hospital, and so on — he explained only half the difference. He was left with what seemed an unsolvable puzzle. "I tried and failed to explain it," he said. "I could have made it more interesting if I had had more confidence." The economics profession responded by rejecting him for every university job he applied for. He settled for a job with a consulting firm.

Then, just as he set out on a new path in life, the firm closed an office and let him go. At the age of twenty-seven, broke and unemployed, with a wife and two little kids, Thaler begged the head of the Rochester School of Management for a job, and the man gave him a temporary one-year gig teaching cost-benefit analysis to business school students. Back in school, he set out to write another dissertation. He found another interesting question: How much is a human life worth? He also found a clever way to approach the problem. He compared the salaries for risky jobs — coal miner, logger, skyscraper window-washer — to the life expectancy of the people who did them. From the data, he backed out what Americans needed to be paid to accept an expected reduction in their life span. If you could calculate what people needed to be paid to accept a 1 percent chance of being killed on the job, you could, in theory, work out what you'd need to pay them to accept a 100 percent chance of being killed on the job. (The number he came up with was $1.4 million, in 2016 dollars.) Later he'd think of his methods as a little silly. ("Do we really think people make this decision rationally?") But older, more successful economists were happy to assume that, say, America's coal miners made some inner calculation of the value of their lives, and charged accordingly.

The paper secured Thaler a full-time job, without tenure, at the Rochester Graduate School of Management. But it was while he was trying to calculate the value of a human life that he began to feel uneasy with economic theory. He'd given questionnaires to subjects that asked them a hypothetical question: If they had been exposed to a virus, and knew there was a one-in-a-thousand chance that they had contracted a fatal disease, how much would they pay for the drug to cure it? Because he was an economist, he knew that there was more than one way to ask the question, so he also asked people: How much would you need to be paid to be exposed to a one-in-a-thousand chance of getting the same fatal disease? Economic theory said the two numbers should be the same. Whatever you were willing to pay to rid yourself of a one-in-a-thousand chance of dying, it should be the same as the sum you needed to be paid to expose yourself to a one-in-a-thousand chance of dying: That number was the value you assigned to a one-in-a-thousand chance of losing your life. People whose lives were even hypothetically on the line didn't feel that way. "The answers people gave were orders of magnitudes apart," said Thaler. "People would say they would pay ten thousand for the cure but would need to be paid a million to be exposed to the virus."

Thaler thought that was really interesting. He told his thesis advisor about his findings. "Stop wasting your time with questionnaires and start doing real economics," said his advisor.

Instead, Thaler began to keep a list. On the list were a lot of irrational things people do that economists claim that they don't do, because economists presume that people are rational. At the top of the list was their willingness to pay 100 times more to avoid a one-in-a-thousand chance of being infected with an incurable disease than they were for the cure for that same disease, after they already had a one-in-a-thousand chance of having it.

Thaler may not have felt all that sure of himself, but he was quick to see that others shouldn't feel so sure of themselves, either. And he noticed that when he had his fellow economists to dinner, they filled up on cashews, which meant they had less appetite for the meal. More to the point, he noticed that they tended to be relieved when he removed the cashew nuts, so they didn't ruin their dinners. "The idea that it could make you better off to reduce your choices — that idea was alien to economics," he said. After he and a friend were given tickets to a basketball game in Buffalo, then decided it wasn't worth driving through a snowstorm to watch it, his friend said, "But if we'd paid for those tickets, we'd be going." An economist would see the tickets as "sunk cost." You don't go to a game you don't want to go to just because you paid for the tickets. Why add to your misery? "I said, 'C'mon, don't you know about sunk cost?' " recalled Thaler. His friend was a computer scientist and didn't know about sunk cost. After Thaler explained the concept, his friend just looked at him and said, "Oh, that's just a bunch of bullshit."

Thaler's list grew quickly. A lot of the items on it fell into a bucket that he eventually would label "The Endowment Effect." The endowment effect was a psychological idea with economic consequences. People attached some strange extra value to whatever they happened to own, simply because they owned it, and so proved surprisingly reluctant to part with their possessions, or endowments, even when trading them made economic sense. But in the beginning, Thaler wasn't thinking in categories. "At the time, I'm just collecting a list of stupid things people do," he said. Why were people so slow to sell vacation homes that, if they hadn't bought them in the first place and were offered them now, they would never buy? Why were NFL teams so reluctant to trade their draft picks when it was obvious that they could often get more than the players were worth in exchange? Why were investors so reluctant to sell stocks that had fallen in value, even when they admitted that they would never buy those stocks at their current market prices? There was no end of things people did that economic theory had trouble explaining. "When you start looking for the endowment effect," Thaler said, "you see it everywhere." His feelings about his own field were not so very different from his feelings for Monopoly as a kid: It was boring, and unnecessarily so. Economics was meant to be the study of an aspect of human nature, but it had ceased to pay attention to human nature. "Thinking about this stuff was *way* more interesting than doing economics," he said.

When he called his observations to the attention of his fellow economists, they weren't interested. "The first thing they'd always say was, 'Of course we know people make mistakes every now and then, but the mistakes are random, and they'll wash out in the market,' " recalled Thaler. Thaler no longer believed that. His list, and the impulse to create it, did not win him friends in the University of Rochester's Department of Economics, or its business school. "He had enemies and he's not awfully good at mollifying enemies," said Tom Russell, a fellow economics professor at Rochester. "If you tell an academic to his face, 'You've just said something really stupid' — okay, the big ones might say, 'How is it stupid?,' but the little ones just store it."

The University of Rochester denied Thaler

tenure. His future was hazy when, in 1976, he attended a conference on how to value a human life. When he heard of Thaler's curious interests, another conference attendee suggested that Thaler read Kahneman and Tversky's article in *Science* that sought to explain why people did stupid things. Thaler went home and found "Judgment Under Uncertainty" in an old copy of *Science.* He couldn't believe his own excitement as he read it. He went and pulled all the other articles in other publications written by Kahneman and Tversky. "I have vivid memories of running from one article to another," says Thaler. "As if I have discovered the secret pot of gold. For a while I wasn't sure why I was so excited. Then I realized: They had one idea. Which was systematic bias." If people could be systematically wrong, their mistakes couldn't be ignored. The irrational behavior of the few would not be offset by the rational behavior of the many. People could be systematically wrong, and so markets could be systematically wrong, too.

Thaler got someone to send him a draft of "Value Theory." He instantly saw it for what it was, a truck packed with psychology that might be driven into inner sanctums of economics and exploded. The logic in the paper was awesome, overpowering. What soon would be known as prospect theory explained most of the items on Thaler's list, in a language economists could understand. There were items on Thaler's list that prospect theory did not address — self-control was the big one — but that didn't matter. The paper blew a hole in economic theory for psychology to enter. "That really is the magic of the paper," said Thaler, "showing you could do it. Math with psychology in it. That paper was what an economist would call proof of existence. It captured so much of human nature."

Till then, Thaler had felt his place in economics to be as uncertain as his ability to copy *Tom Sawyer.* "If they didn't exist, I don't know if I would have stayed in the field," he said. After finishing the collected works of the Israeli psychologists, he had a new feeling. "The way it feels to me," he said, "is that there were certain ideas that I was put on this earth to think. And now I can think them." He would begin, he decided, by turning his list into an article. But even before he did, he found a mailing address for the Department of Psychology at Hebrew University and wrote a letter to Amos Tversky.

■ ■ ■ ■

It was almost always to Amos that the economists wrote. They understood him. Amos's insistently logical mind was much like their own but somehow better: They could see his genius. To most economists, Danny's mind was a mystery. Richard Zeckhauser, an economist at Harvard who became friends with Amos, spoke for his entire field when he said, "My impression of the way they worked on a paper is that they walked around and let Danny do a variety of things. 'Guess what, Amos, I went to buy a car and I offered 38 grand and the salesman said 38.9 and I said yes! Did I do a good job?' And Amos would say, 'Let's write that up.' " The economists' view of the collaboration was that Amos Tversky had set out, like an anthropologist, to study an alien tribe of beings less rational than himself, and his tribe was Danny. "I share your feeling that such behavior is, in some sense, unwise or erroneous, but this does not mean that it does not occur," Amos wrote, to an American economist who complained about the description of human nature implied by "Value Theory." "A theory of vision cannot be faulted for predicting optical illusions.

Similarly, a descriptive theory of choice cannot be rejected on the grounds that it predicts 'irrational behavior' if the behavior in question is, in fact, observed."

Danny, for his part, claimed that it wasn't until 1976 that he woke up to the effects their theory might have on a field he knew nothing about. His awakening came when Amos handed him a paper written by an economist. The paper opened, "The agent of economic theory is rational, selfish, and his tastes do not change." The economists at Hebrew University were in the building next door, but Danny hadn't paid any attention to their assumptions about human nature. "To me, the idea that they really believed in it — that this is really their worldview — was incredible," he said. "It's the worldview in which if people tip in a restaurant to which they will never return it counts as a puzzle." It was a worldview that took it as given that the only way to change people's behavior was to change their financial incentives. The idea of it struck him as so bizarre that he could scarcely bring himself to engage with it directly. To Danny the whole idea of proving that people weren't rational felt a bit like proving that people didn't have fur. Obviously people were not rational, in any meaningful sense of that term.

He and Amos wanted to avoid getting into an argument about the rationality of man. That argument would only distract people from the phenomenon they were uncovering. They preferred to reveal man's nature, and let man decide what he was. Their next task, they saw, was to buff and polish "Value Theory" for publication. They both worried that someone would find an obvious contradiction — some Allais paradox–like observation that would render their theory dead on arrival. They'd spend three years doing very little else but searching the theory for internal contradictions. "In those three years we did not discuss anything of genuine interest," said Danny. Danny's interest ended with the psychological insights; Amos was obsessed with the business of using the insights to create a structure. What Amos saw, perhaps more clearly than Danny, was that the only way to force the world to grapple with their insights into human nature was to embed them in a theory. That theory needed to explain and predict behavior better than existing theory, but it also needed to be expressed in symbolic logic. "What made the theory important and what made it viable were completely different," said Danny, years later. "Science is a conversation and you have to compete for the right to be heard. And the competition has its rules. And the rules, oddly enough, are that you are tested on formal theory." After they finally sent a draft of their paper to the economics journal *Econometrica,* Danny was perplexed by the editor's response. "I was kind of hoping he'd say, 'Loss aversion is a really cool idea.' He said, 'No, I like the math.' I was sort of shattered."

By 1976, purely for marketing purposes, they changed their title to "Prospect Theory." "The idea was to give the theory a completely distinct name that would have no associations whatsoever," said Danny. "When you say 'prospect theory,' no one knows what you're talking about. We thought: Who knows? It may turn out to be influential. And if it is we don't want it to be confused with anything else."

In all of this they were slowed, dramatically, by the turmoil in Danny's life. By 1974 he'd moved out of his house and was living apart from his wife and children. A year later he left the marriage, and flew to London to meet the psychologist Anne Treisman to formally "declare my love." She reciprocated. By the fall of 1975 Amos was clearly weary of the inevitable fallout. "It is hard to overestimate the amount of time and the amount of emotional and mental energy that is consumed with such affairs," he wrote to his friend Paul Slovic.

In October 1975 Danny flew to England again, this time to see Anne in Cambridge and to travel with her to Paris. He was at once in a totally uncharacteristic state of elation and worried about the effect of his new relationship with Anne on his old one with Amos. In Paris he found waiting what appeared to be a letter from Amos — but, opening it, he at first found only a draft of what would become "Prospect Theory." Danny took the absence of any personal note as a subtle message from Amos. Sitting with his new love in the world's capital of romance, Danny sat down and wrote what amounted to a love letter: to Amos. "Dear Amos," it began. "When I came to Paris I found an envelope from you. I pulled out your manuscript but there was no letter with it. And I told myself that Amos is very angry with me, and not without reason. After dinner, I was looking for a used envelope to send this back to you and I found your envelope, and then saw your letter inside. We were late for dinner and I just glanced to see how you finish it. And I saw the words 'Yours, as ever' and I had goose bumps from emotion." He went on to write that he'd explained to Anne that he could never have achieved on his own what he had achieved with Amos, and that the new paper they were working on was yet another step. "This is for me the greatest moment in a relationship which I see as one of the peaks of my life," he wrote. Then he added: "I was yesterday at Cambridge. And I spoke to them about our work on Value Theory. The enthusiasm is almost embarrassing. I concluded with a discussion of the early stages of the isolation effect. And they responded to that especially. In general, they gave me the feeling that I'm one of the world's greats. They were trying so hard to impress me that I reached the conclusion that maybe the time has come for me to be free of the need to impress others."

In some strange way, as they approached their moment of greatest public triumph, their collaboration remained a private affair, a gamble with no context. "As long as we stayed in Israel, the whole idea of what the world thought of us didn't occur to us," said Danny. "We benefited from our isolation." That isolation depended on them being together, in the same room, behind a closed door.

That door was now cracking open. Anne was British. She was also a gentile and the mother of four children, one of whom had Down syndrome. There were about sixteen different reasons she couldn't, or wouldn't, move to Israel. And if Anne wasn't moving to Israel, it followed that Danny would need to leave. Danny and Amos scrambled and found a temporary solution, in 1977, by heading off together from Hebrew University on sabbatical to Stanford University, where Anne might join them. But a few months after their arrival in the United States, Danny announced that he planned to marry Anne and stay. He forced Amos to make a decision about what to do about their relationship.

It was now Amos's turn to sit down and write an emotional letter. Danny was messy, in a way that Amos could never be messy even if he wanted to be. Amos had wanted to be a poet when he was a boy. He'd wound up a scientist. Danny *was* a poet, who somehow happened to have become a scientist. Danny felt some obvious desire to be more like Amos; Amos, too, harbored some less obvious desire to be more like Danny. Amos was a genius. But he needed Danny, and he knew it. The letter Amos wrote was to his close friend Gidon Czapski, the rector of Hebrew University. "Dear Gidi," it began. "The decision to remain here in the United States is the most difficult decision I have ever made. I cannot ignore my desire to bring to a completion, at least partially, the joint work with Danny. I just cannot accept the idea that the joint work of years could come to naught and that we will not be able to complete the ideas we have." Amos went on to explain that he planned to accept a chaired professorship offered to him by Stanford University. He knew full well that everyone in Israel would be shocked and angry. "If Danny leaves Israel it is a personal tragedy," a Hebrew University official had said to him not long before. "If you leave it is a national tragedy."

Until Amos actually left, his friends found it unthinkable that he would live anyplace but Israel. Amos was Israel, and Israel was Amos. Even his American wife was upset. Barbara had fallen in love with Israel — its intensity, its sense of community, its disinterest in small talk. She now thought of herself as more Israeli than American. "I had done so much work to become Israeli," she said. "I didn't want to stay in the States. I said to Amos, 'How can I start over?' He said, 'You'll manage.' "

11
THE RULES OF UNDOING

In the late 1970s, not long after he'd become superintendent of the Massachusetts Mental Health Center, Miles Shore realized he had a problem. The center was a teaching hospital for Harvard Medical School, where Shore was Bullard Professor of Psychiatry. Newly installed in administration, he found himself faced with a decision: whether to promote a medical researcher named J. Allan Hobson. It shouldn't have been that hard. In a series of famous papers, Hobson had landed body blows on the Freudian idea that dreams arose from unconscious desires, by showing that they actually came from a part of the brain that had nothing to do with desire. He'd proven that the timing and the length of dreams were regular and predictable, which suggested that dreams had less to say about a person's psychological state than about his nervous system. Among other things, Hobson's research suggested that people who paid psychoanalysts to find meaning in their unconscious states were wasting their money.

Hobson was changing people's understanding of what happened to the human brain during sleep — but he wasn't doing it alone. That was Miles Shore's problem: Hobson hadn't written his famous papers on dreams by himself, but with a partner named Robert McCarley. "It was very difficult to campaign for promotion for people who did their work collaboratively," said Shore. "Because the system is based on the individual. It was always: What did this person do to change the field?" Shore wanted to promote Hobson, but he had to argue the case before a skeptical committee. "They basically didn't want to promote anyone," said Shore. Resisting the case for Hobson, committee members asked Shore if he could demonstrate exactly how much Hobson had contributed to his partnership with McCarley. "They asked me which one of them did what," recalled Shore. "And so I went to them [Hobson and McCarley] and asked: 'Which one of you did what?' And they said: 'Which one of us did what? We have no idea. It was a joint product.' " Shore pushed the collaborators a bit until he realized that they really meant it: They had no idea who deserved credit for which idea. "It was really interesting," said Shore.

So interesting that Shore decided there might be a book in it. He set out to find fertile pairs — people who had been together for at least five years and produced interesting work. By the time he was done he had interviewed a comedy duo; two concert pianists who had started performing together because one of them had stage fright; two women who wrote mysteries under the name "Emma Lathen"; and a famous pair of British nutritionists, McCance and Widdowson, who were so tightly linked that they'd dropped their first names from the jackets of their books. "They were very huffy about the idea that dark bread was more nutritious than white bread," recalled Shore. "They had produced the research that it wasn't so in 1934 — so why didn't people stop fooling around with the idea?" Just about every work couple that Shore called were intrigued enough by their own relationships to want to talk about them. The only exceptions were "a mean pair of physicists" and, after flirting with participating, the British ice dancers Torvill and Dean. Among those who agreed to sit down with Miles Shore were Amos Tversky and Daniel Kahneman.

Shore found Danny and Amos together in August 1983, in Anaheim, California, where they were attending the American Psychological Association meeting. Danny was now forty-nine and Amos forty-six. They spoke with Shore together for several hours and then, for several hours more, separately. They walked Miles Shore through the history of the collaboration, starting with their early excitement. "In the beginning we were able to answer a question that had not been asked," Amos told him. "We were able to take psychology out of the contrived laboratory and address the topic from the experiences all around us." Trying to pin them down on the question they thought they were answering, Shore asked if their work fed into the new and growing field of artificial intelligence. "You know, not really," said Amos. "We study natural stupidity instead of artificial intelligence."

The Harvard psychiatrist thought that Danny and Amos had a lot in common with other successful pairs. The way they had created what amounted to an exclusive private club of two, for example. "They were crazy about each other, and not indiscriminate," said Shore. "They were not generally crazy about other people. They *hated* editors." As with some of the other fertile pairs, the partnership had created strains on their other close relationships. "The collaboration has put a lot of pressure on my marriage," Danny confessed. Like the other pairs, they had lost any sense of individual contribution. "You ask who did what?" said Danny "We didn't know at that time, not clearly. It was beautiful, not knowing." Shore thought that both Amos and Danny realized, or seemed to realize, how much they needed each other. "There are geniuses who work on their own," said Danny. "I am not a genius. Neither is Tversky. Together we are exceptional."

What set Amos and Danny apart from the nineteen other couples Shore had interviewed for his book was their willingness to speak about the problems in their relationship. "When I asked about conflicts, most people just ignored it," said Shore. "A number didn't want to admit there was any conflict." Not Amos and Danny. Or, at any rate, not Danny. "It's been difficult since I got married and since we moved to this continent," he confessed. Amos remained evasive, and yet great chunks of Shore's conversation with Danny and Amos wound up being about the many troubles they'd had since leaving Israel six years earlier.

With Amos in the room, Danny complained at length about how different the public perception of the collaboration was from its reality. "I am perceived as attending him, which is not the case," he said, less to Shore than to Amos. "I clearly lose by the collaboration. There is a quality that is clearly contributed by you. Formal analysis is not my strength and it shows up very distinctly in our work. My contributions are less unique." Amos spoke, at less length, about how the blame for their unequal status fell squarely on other people. "The credit business is very hard," said Amos. "There is a lot of wear and tear, and the outside world isn't helpful to collaborations. There is constant poking, and people decide that one person gets the short end of the stick. It's one of the rules of balance, and joint collaboration is an unbalanced structure. It is just not a stable structure. People aren't happy with it."

Alone with the Harvard psychiatrist, Danny said more. He hinted that he didn't believe the outside world was entirely responsible for the problems in their relationship. "The spoils of academic success, such as they are — eventually one person gets all of it, or gets a lot of it," he said. "That's an unkindness built in. Tversky cannot control this, though I wonder whether he does as much to control it as he should." Then he came straight out with his own feelings about Amos getting the lion's share of the glory for work they had done together. "I am very much in his shadow in a way that is not representative of our interaction," he said. "It induces a certain strain. There is envy! It's just disturbing. I *hate* the feeling of envy. . . . I am maybe saying too much now."

Shore left the interview feeling that Amos and Danny had just come through a rough patch, but that the worst lay behind them. Their openness about their problems he took as a good sign. They hadn't exactly been fighting during their interview; their attitude toward conflict was just different than that of the other couples he had spoken with. "They played the Israeli card," said Shore. *"We're Israeli, so we yell at each other."* Amos, especially, sounded optimistic that he and Danny would continue to work together as much as they had. It helped, Danny and Amos agreed, that the American Psychological Association had just honored both of them with its Distinguished Scientific Contribution Award. "I have lived in some fear that he might get it alone," Danny confessed to Shore. "That would have been a disaster, and I couldn't have coped very elegantly." The award had eased some pain. Or so it seemed to Miles Shore.

As it happened, Shore never wrote his book about fertile pairs. But years later, he sent Danny an audiotape of their conversation. "I listened to it," Danny said, "and it is absolutely clear from it that we are finished."

In late 1977, after Danny had told him that he wasn't returning to Israel, word spread through academia that Amos Tversky might leave, too. The job market for college professors typically moves slowly and with great reluctance, but in this instance it leapt into action. It was as if an especially deliberate fat man watching TV on his couch suddenly realized that his house was on fire. Harvard University quickly offered Amos tenure, though it took them a few weeks to throw in an assistant professorship for Barbara. The University of Michigan, which had the advantage of sheer size, scrambled to find four tenured professorships — and, by making places for Danny, Anne, and Barbara, also snag Amos. The University of California at Berkeley, which left Danny with the clear impression, when he made overtures, that he was too old to be hired, prepared to offer a job to Amos. But no place moved quite so dramatically as Stanford.

The psychologist Lee Ross, a rising young star on the Stanford faculty, led the charge. He knew that the big public American universities who wanted Amos might, in the bargain, offer jobs to Barbara and Danny and Anne. Stanford was smaller and didn't have four jobs to offer. "We figured there were two things we could do that those schools might not," said Ross. "One was to make the offer early, and the other was to make it fast. We wanted to convince him to come to Stanford, and the best way we can convince him to come is to show him how quickly we can act."

What happened next was, Ross believed, unprecedented in the history of the American university. The morning he learned Amos was on the market, he convened Stanford's Psychology Department. "I was supposed to present the case for Amos," said Ross. "I said, I'm going to tell you a classic Yiddish story. There's a guy, an eligible bachelor. A happy bachelor. The matchmaker comes to him and says, 'Listen, I have for you a match.' 'Ah, I'm not so sure,' says the bachelor. 'She's really special,' says the matchmaker. 'What, is she beautiful?' asks the bachelor. 'Beautiful? She looks like Sophia Loren, only younger.' 'What, does she have family money?' asks the bachelor. 'Money? She's an heiress to the Rothschild fortune.' 'Then she must be a dope,' says the bachelor. 'A dope? She has been nominated for Nobel Prizes in both physics and chemistry.' 'I accept!' says the bachelor. To which the matchmaker replies, 'Good, we have half a match!' " Ross told the Stanford faculty, "After I tell you about Amos, you will say, 'I accept!' and I will say, 'I'm sorry to tell you we have half a match.' "

Even to Ross it was unclear that the sales pitch was necessary. "Everyone who came across the work congratulated themselves on their own good judgment and insight in appreciating the work," said Ross. "But nobody didn't get it." That same day, the Stanford Psychology Department went to the Stanford president and said: *We have none of the usual paperwork. No recommendations or anything else. Just trust us.* Stanford made Amos an offer of lifetime employment that afternoon.

Amos would later tell people that in choosing between Harvard and Stanford, he imagined the regret he would experience at each. At Harvard he'd regret passing up Palo Alto's weather and living conditions, and resent the commute; at Stanford he'd regret, and only briefly, not being able to say he was a Harvard professor. If it occurred to him or anyone else that Amos, to be Amos, needed Danny close at hand, he didn't show it. Stanford showed not the slightest interest in Danny. "There's a practical issue," said Ross. "Do you want two guys doing the same thing? And the cold fact is we got the full benefit of Danny and Amos just by hiring Amos." Danny would have loved for them all to go to Michigan, but Amos clearly had no interest in anyplace but Harvard or Stanford. After Harvard and Stanford had ignored him, and Berkeley had let him know that he would not be offered a job, Danny accepted a position beside Anne at the University of British Columbia, in Vancouver. He and Amos agreed they would take turns flying to visit each other every other weekend.

Danny was still floating on air. "We were on such a high from having finished prospect theory and embarked on framing that we must have felt pretty invulnerable," he said. "There was not a shadow between us at the time." He watched Amos give the traditional job-application talk at Stanford, *after* Stanford had made him what was likely the fastest job offer in its history.

Amos presented prospect theory. "I noticed that I felt nothing but pride for him," said Danny. "I noticed it because envy would have been natural." When Danny left Palo Alto for Vancouver for the start of the 1978–79 academic year, he was even more aware than usual of the serendipity of life. His two children were now on the other side of the world, along with his old lab, a department full of former colleagues, and a society to which he once assumed he belonged. He had left behind in Israel a ghost of himself. "The background to what I was thinking was that I had just changed my life," he said. "I'd changed my wife. The counterfactuals were with me all the time. I was constantly comparing my life to what it might have been."

In this curious state of mind, he found his thoughts settling on a nephew, Ilan. Ilan had been a twenty-one-year-old navigator in the back of an Israeli fighter during the Yom Kippur war. After the war, he had sought out Danny and asked him to listen to an audiotape he had kept from it. He'd been in the backseat of the fighter when an Egyptian MiG got behind them, locking in for a kill. On the tape, you could hear Ilan scream at his pilot, "Break! Break! Break! He's on our tail!" As Ilan played the tape,

Danny noticed that the young man was shaking; for some reason, he wanted his uncle to hear what had happened to him. Ilan had survived the war, but a year and a half later, in March 1975, five days before he was to be released from service, he was killed. Blinded by a flare, his pilot had flown upside down straight into the ground.

They'd thought they were rising when in fact they were falling. It wasn't an original mistake. Pilots in flight often became disoriented. The inner ear wasn't designed for a gravity-defying chamber pitching and rolling at 650 miles an hour a mile above the earth's surface any more than the mind was designed to calculate the probabilities of complex situations. It had evolved to stabilize people on their own two feet. People who flew airplanes were susceptible to sensory illusions — which was why a pilot without an instrument rating who flew into clouds had an average life expectancy of 178 seconds.*

After Ilan's death, Danny couldn't help but notice the urge in those who loved him to mentally undo his plane crash. Many of

* That strange fact comes from an excellent article on the subject of pilot illusions by Tom LeCompte in the Smithsonian's *Air & Space* magazine.

the sentences that came from their lips might just as well have started with the words "if only." *If only* Ilan had been released from the Air Force a week earlier. *If only* he'd taken charge after his pilot was blinded by that flare. People's minds coped with loss by drifting onto fantasy paths, where loss never occurred. But this drifting, Danny noticed, wasn't random. There appeared to be constraints on the mind when it created alternatives to reality. If Ilan had still had a year of service remaining when his plane crashed, no one would have said, "If only he had been released a year ago." No one said, "If only the pilot had the flu that day" or "If only Ilan's plane had been grounded for mechanical problems." For that matter, no one said, "If only Israel had not had an Air Force." Any of those counterfactuals would have saved his life, but none of them came to the minds of those who loved him.

There were of course a million ways that any plane crash might have been avoided, but people seemed to consider only a few of them. There were patterns in the fantasies that people created to undo his nephew's tragedy, and they resembled patterns in the alternative versions of his own life that played out in Danny's mind.

Soon after his arrival in Vancouver, Danny asked Amos to send him any notes that he'd kept from their discussions about regret. In Jerusalem they'd spent more than a year talking about the rules of regret. They'd been interested chiefly in people's anticipation of the unpleasant emotion, and how this anticipation might alter the choices they made. Now Danny wanted to explore regret, and other emotions, from the opposite direction. He wanted to study how people undid events that had already happened. Both he and Amos could see how such a study might feed into their work on both judgment and decision making. "There is nothing in the framework of decision theory that would prohibit the assignment of utilities to states of frustrated hope, relief or regret, if these are identified as important aspects of the experience of consequences," they wrote, in what amounted to a memo to themselves. "However, there is a reason to suspect a major bias against the acknowledgment of the true impact of such states on experience. . . . It is expected of mature individuals that they should feel the pain or pleasure that is appropriate to the circumstances without undue contamination by unrealized possibilities."

Danny now had an idea that there might

be a fourth heuristic — to add to availability, representativeness, and anchoring. "The simulation heuristic," he'd eventually call it, and it was all about the power of unrealized possibilities to contaminate people's minds. As they moved through the world, people ran simulations of the future. What if I say what I think instead of pretending to agree? What if they hit it to me and the grounder goes through my legs? What happens if I say no to his proposal instead of yes? They based their judgments and decisions in part on these imagined scenarios. And yet not all scenarios were equally easy to imagine; they were constrained, much in the way that people's minds seemed constrained when they "undid" some tragedy. Discover the mental rules that the mind obeyed when it undid events after they had occurred and you might find, in the bargain, how it simulated reality before it occurred.

Alone in Vancouver, Danny was gripped by his new interest in the distance between worlds — the world that existed and worlds that might have come to pass but never did. Much of the work he and Amos had done was about finding structure where no one had ever thought to look for it. Here was another chance to do that. He wanted to investigate how people created alternatives to reality by undoing reality. He wanted, in short, to discover the rules of the imagination.

With one eye on a prickly colleague in his new department named Richard Tees, Danny sat down and created a vignette for a new experiment:

> Mr. Crane and Mr. Tees are scheduled to leave the airport on different flights, at the same time. They traveled from town in the same limousine, were caught in the same traffic jam, and arrived at the airport thirty minutes after the scheduled departure time of their flights.
>
> Mr. Crane is told that his flight left on time.
>
> Mr. Tees is told that his flight was delayed, and just left five minutes ago.
>
> Who is more upset?

The situation of the two men was identical. Both expected to miss their planes and both had. And yet 96 percent of the subjects to whom Danny put the question said that Mr. Tees was more upset. Everyone seemed to understand that reality wasn't the only source of frustration. The emotion was also fed by its proximity to another reality — how "close" Mr. Tees came to making his flight. "The only reason for Mr. Tees to be more upset is that it was more 'possible' for him to reach his flight," Danny wrote, in notes for a talk on the subject. "There is an Alice-in-Wonderland quality to such examples, with their odd mixture of fantasy and reality. If Mr. Crane is capable of imagining unicorns — and we expect he is — why does he find it relatively difficult to imagine himself avoiding a thirty-minute delay, as we suggest he does? Evidently there are constraints on the freedom of fantasy."

It was those constraints that Danny set out to investigate. He wanted to understand better what he was now calling "counterfactual emotions," or the feelings that spurred people's minds to spin alternative realities in order to avoid the pain of the emotion. Regret was the most obvious counterfactual emotion, but frustration and envy shared regret's essential trait. "The emotions of unrealized possibility," Danny called them, in a letter to Amos. These emotions could be described using simple math. Their intensity, Danny wrote, was a product of two variables: "the desirability of the alternative" and "the possibility of the alternative." Experiences that led to regret and frustration were not always easy to undo. Frustrated people needed to undo some feature of their environment, while regretful people needed to undo their own actions. "The basic rules of undoing, however, apply alike to frustration and regret," he wrote. "They require a more or less plausible path leading to the alternative state."

Envy was different. Envy did not require a person to exert the slightest effort to imagine a path to the alternative state. "The availability of the alternative appears to be controlled by a relation of similarity between oneself and the target of envy. To experience envy, it is sufficient to have a vivid image of oneself in another person's shoes; it is not necessary to have a plausible scenario of how one came to occupy those shoes." Envy, in some strange way, required no imagination.

Danny spent the first several months of his separation from Amos with these strange and beguiling thoughts. In early January 1979, he wrote Amos a memo titled "The state of the 'undoing' project." "I have spent some time making up disasters and undoing them in various ways," he wrote, "in an attempt to order the alternative modes of undoing."

> A shopkeeper was robbed at night. He resisted. Was beaten in the head. Was left alone. Eventually died before robbery was noticed.

> A head-on collision between two cars, each attempting to overtake under conditions of restricted visibility.

> A man had a heart attack, tried in vain to reach the phone.

> Someone is killed by a stray shot in a hunting accident.

"How do you undo those?" he wrote. "And Kennedy's assassination. World War II?" He went on for eight neatly written pages. Imagination wasn't a flight with limitless destinations. It was a tool for making sense of a world of infinite possibilities by reducing them. The imagination obeyed rules: the rules of undoing. One rule was that the more items there were to undo in order to create some alternative reality, the less likely the mind was to undo them. People seemed less likely to undo someone being killed by a massive earthquake than they were to undo a person's being killed by a bolt of lightning, because undoing the earthquake required them to undo all the earthquake had done. "The more consequences an event has, the larger the change that is involved in eliminating that event," Danny wrote to Amos. Another, related, rule was that "an event becomes gradually less changeable as it recedes into the past." With the passage of time, the consequences of any event accumulated, and left more to undo. And the more there is to undo, the less likely the mind is to even try. This was perhaps one way time heals wounds, by making them feel less avoidable.

A more general rule Danny labeled "The Focus Rule." "We tend to have a hero or an actor operating in a situation," he wrote. "Wherever possible we'll keep the situation fixed and have the actor move. . . . We don't invent a gust of wind to deflect Oswald's bullet." An exception to this rule was when the person engaged in the undoing was the main actor of his own fantasy. He was less likely to undo his own actions than he was to undo the situation in which he found himself. "Changing or replacing oneself is much less available than changing or replacing another actor," wrote Danny. "A world in which I have a new set of traits must be very far from the world in which I live. I may have <u>some</u> freedom, but I am not free to be someone else."

The most important general rule of undoing had to do with what was surprising or unexpected. A middle-aged banker takes the same route to work every day. One day he takes a different route and is killed when a drugged-out kid in a pickup truck runs a red light and sideswipes his car. Ask people to undo the tragedy, and their minds drift to the route the banker took that day. If only he had gone the usual way! But put that same man back on his normal route, and let him be killed by the same drugged-out boy in the same truck, running a different stoplight, and no one thought: If only he had taken a different route that day! The distance the mind needed to travel from the usual way of doing things to some less ordinary way of doing things felt further than the trip made from the other direction.

In undoing some event, the mind tended to remove whatever felt surprising or unexpected — which was different from saying that it was obeying the rules of probability. A far more likely way to spare the man was to alter his timing. If he or the boy had been just a few seconds faster or slower at any moment on their tragic journeys, they'd never have collided. When undoing the accident, people didn't think of that. It was easier to undo the unusual part of the story. "You may amuse yourself by mentally undoing Hitler," Danny wrote, then mentioned to Amos a recent history that imagined Hitler having succeeded in his original ambition, to be a painter in Vienna. "Now imagine another [counterfactual]," wrote Danny. "Simply remember that just prior to the instant of conception there was a better than even chance that Adolf Hitler would be a lady. The probability of his being a successful artist was perhaps never so high [as the better than 50-50 chance that he would be born a girl]. Why then do we find one of these approaches to undoing Hitler quite acceptable and the other shocking, almost ungrammatical?"

The workings of the imagination reminded Danny of cross-country skiing, which he'd tried and failed to take up in Vancouver. He'd taken the beginner's course twice, and discovered mainly how much more effort it took to climb some hill than to ski down it. The mind also preferred to go downhill when it was engaged in undoing. "The Downhill Rule," Danny called this.

As he worked out this new idea, he had a

new feeling — of having gone fast and far without Amos. At the end of his letter, he wrote, "It would help a lot if you could spend a couple of hours writing *me* a letter about this, before we meet next Sunday." Danny wouldn't recall if Amos ever wrote that letter — most likely he didn't. Amos seemed interested in Danny's new ideas, but for some reason he didn't contribute to them. "He had little to say, which was rare for Amos," said Danny. He suspected that Amos was wrestling with unhappiness, which was also unlike Amos. After he left Israel, Amos would later confide in a close friend, he was surprised by how little guilt he felt, and also by how much homesickness. Maybe that was the problem; maybe Amos, having formally immigrated to the United States, wasn't feeling himself. Or maybe the problem was how different these new ideas felt from their other work. Their work until then had always started as a challenge to some existing, widely accepted theory. They exposed the flaws in theories of human behavior and created other, more persuasive theories. There was no general theory of the human imagination to disprove. There was nothing to destroy, or really even to push up against.

There was another problem — the dramatic new difference in their relative status was coming between them. When Amos visited the University of British Columbia, he seemed to be lowering himself. Danny *went up* to Palo Alto and Amos *came down* to Vancouver. "Amos was a contemptuous person, and I could sense how provincial he thought the place was," said Danny. One night as they talked, Amos blurted out that the difference he felt being at Stanford was the difference of being in a place where everyone was first-rate. "That was a first," recalled Danny. "I knew he really did not mean a thing by it and that he probably regretted saying it — but I remember the thought that it was simply inevitable that Amos would feel some condescending pity and that I would be hurt by it."

But Danny's overwhelming feeling was of frustration. He'd gone the better part of a decade having all his ideas more or less in Amos's presence. There was no time at all between the moment either of them had some idea and the moment he shared it with the other. The magic was what happened next: the uncritical acceptance, the joining together of their minds. "I have a feeling that I initiate a lot, but the product is always out of my reach," Danny would one day tell Miles Shore. Now he was back to working alone, sensing an absence of thoughts that would improve his own. "I was having an enormous number of ideas, but he wasn't there," said Danny. "And so those ideas were wasted, because they didn't have the benefit of the kind of thinking that Amos was capable of putting into things."

A few months after Danny wrote his memo to Amos, in April 1979, he and Amos delivered a pair of talks at the University of Michigan. The occasion was the prestigious annual Katz-Newcomb Lecture Series, and the striking thing about it, to Danny, was that both of them had been invited, not just Amos. Danny's impression that Amos might be feeling low on new ideas was confirmed when Amos took for the subject of his talk their joint work on framing. Danny's was his first public unveiling of ideas he had cooked up in their nine months apart. "The Psychology of Possible Worlds," he called it. "Because we feel ourselves to be among friends," he began, "Amos and I have decided on what otherwise would be a risky choice for this lecture. A topic that we have only recently begun to study, and about which we still have much more enthusiasm than we have knowledge. . . . We shall explore the role of unrealized possibilities in our emotional response to reality and in our understanding of it."

He then explained the rules of undoing. He had created more vignettes to test on subjects — in addition to a banker who was killed in a car crash by a drugged-out boy, there was another unlucky man, who had died of both a heart attack and from failing to hit the brakes on his car. Most of them he'd dreamed up late at night in Vancouver. He'd been awakened so often by his thoughts on the subject that he'd kept a notepad by his bed. Amos might be the superior mind, but Danny was the better talker. Amos might be getting the lion's share of the rewards of their relocation to North America, but that couldn't last forever: People would see his contribution. The audience was enthralled — he could see that. And when he was done, no one was in a hurry to leave. They were standing around together, and Amos's old mentor Clyde Coombs approached them with genuine wonder in his eyes. "The ideas, so many ideas, where do they come from?" he asked. And Amos said, "Danny and I don't talk about these things."

Danny and I don't talk about these things.

That was the moment when the story unspooling in Danny's mind began to change. Later he would point to it and say: That is the beginning of the end of us. He would later seek to undo the moment, but when he did, he did not say, "If only Clyde Coombs had not asked that question." Or: "If only I felt as invulnerable as Amos." Or: "If only I had never left Israel." He said, "If only Amos was capable of self-effacement." Amos was the actor in Danny's imagination. Amos was the object in focus. Amos had been handed on a platter a chance to give Danny credit for what he had done, and Amos had declined to take it. They'd move on, but the moment had lodged itself in Danny's mind and would refuse to leave it. "Something happens when you are with a woman you love," said Danny. "You know something happened. You know it's not good. But you go on." You are in love, and yet you sense a new force pulling you out of it. Your mind has lit upon the possibility of another narrative. You half hope something comes along to stabilize or reenergize the old one. In this case, nothing came along. "I wanted Amos to lean back against what was happening and he was not doing it, nor did he accept that he had to do it," said Danny.

After Michigan, Danny gave talks about the undoing project and neglected to mention Amos. He'd never done that sort of thing before. For a decade, they had had a hard-and-fast rule against inviting others anywhere near areas of mutual interest. At the end of 1979, or perhaps in early 1980, Danny began to talk to a young assistant professor at UBC named Dale Miller, sharing his ideas about the way people compared reality to its alternatives. When Miller asked about Amos, Danny said that they were no longer working together. "He was in Amos's shadow and he was very worried about that, I think," said Miller. It wasn't long before Danny and Miller were working on a paper together that might just as well have been called "The Undoing Project." "I thought that they had agreed to see other people," said Miller. "And he was insistent that his days of collaborating with Amos were over. I remember a lot of fraught conversations. At some point he said to be gentle with him, because this was his first relationship after Amos."

If the Katz-Newcomb Lecture meant less to Amos than to Danny, it was because Amos's life was now a sprint from one Katz-Newcomb Lecture to the next. He reminded at least one of his new graduate students at Stanford of a stand-up comic, traveling the world and working small nightclubs to test his material. "He thought by talking," recalled his wife, Barbara. "You could hear him in the shower. You could hear him talking to himself. Through the door." His children grew used to hearing their father alone in a room, talking. "It was a little bit like an insane person talking to himself," said his son Tal. They'd see him coming home in his brown Honda, stopping and starting in the street in front of their house and talking. "He'd be going three miles an hour, then all of a sudden he'd gun it," said his daughter, Dona. "He'd worked out the idea."

In the weeks leading up to the Katz-Newcomb Lecture, in early April 1979, Amos was busy talking to the Soviet Union. He'd joined a delegation of ten prominent Western psychologists on a bizarre intellectual diplomatic mission. Soviet psychologists were then trying to persuade their government to admit mathematical psychology into the Russian Academy of Sciences and had asked their American counterparts for support. Two distinguished mathematical psychologists, William Estes and Duncan Luce, had taken it upon themselves to help them. The older guys made a short list of America's leading mathematical psychologists. Most of them were ancient. Amos counted as one of the younger guys, along with his Stanford colleague Brian Wandell. "The older guys had this idea that we were going to rescue the image of psychology in the Soviet Union," recalled Wandell. "Psychology flew in the face of Marxism. It was on the list of things that didn't need to exist."

It took about a day to realize why Marxism might feel that way. These particular Soviet psychologists were charlatans. "We were thinking there really were going to be scientists on the Soviet end," said Wandell. "There weren't." The Soviets and the Americans took turns giving presentations. An American would give a learned talk about decision theory. His Soviet counterpart would rise and offer a talk that sounded completely insane — one guy spent his allotted time on his theory about how the brain waves caused by beer canceled the brain waves caused by vodka. "We'd get up and give a paper, and you know, it was okay," said Wandell. "Then some Russian guy would get up and talk and we'd say, 'Wow, that was weird.' One was about how the meaning of life could be put into a formula and the formula might have some variable labeled *E* in it."

With one exception, the Russians knew

nothing about decision theory, and didn't even seem particularly interested in the subject. "There was one guy," said Wandell, "who gave this great talk, at least compared to the others." That guy turned out to be a KGB agent, whose training in psychology consisted of the talk he had given. "The way we discovered he was a KGB guy was that he showed up later at a physics conference and gave a great talk there, too," said Wandell. "*That* was the only guy Amos liked."

They stayed in a hotel where the toilets didn't flush and the heat didn't work. Their rooms were bugged, and everywhere they went they were followed by guards. "People were pretty freaked out the first day or two," said Wandell. "We were plainly in over our heads." Amos found the whole thing hysterical. "They put a focus on Amos, probably because he was Israeli," said Wandell. "In typical Amos fashion, he was walking around Red Square, and gives me this look that says, 'C'mon, let's lose 'em!' Then he just kind of took off, with the guards chasing after him." When they finally caught up to him — hiding in a department store — the Soviets were furious. "They gave us all a stern talking-to," said Wandell.

Amos spent at least some of his time in his bugged and heatless hotel room adding to a file that he'd labeled "The Undoing Project." The file in the end came to forty or so pages of handwritten notes. Between the lines, you can hear the polite throat clearing of a diamond cutter waiting for his rocks. Amos clearly had hopes of turning Danny's ideas into a full-blown theory. Danny didn't know that, or that Amos was busy dreaming up his own vignettes:

> David P was killed in a plane crash. Which of the following is easier to imagine:
>
> ________ that the plane did not crash
> ________ that David P. took another plane

Instead of replying to Danny's long letter, Amos made notes to himself, trying to order the stuff spilling out of Danny. "The present world is often surprising, i.e., less plausible than some of its alternatives," he wrote. "We can order possible worlds by i) initial plausibility and ii) similarity to the present world." He followed this a few days later with eight dense pages in which he attempted to create a logical, internally consistent theory of the imagination. "He loved these ideas," said Barbara. "It's something very basic about decision making that fascinated him. It's the choice you don't take." He groped for titles, so that he might know what he was writing about. In his earliest notes in the file, he scribbled the phrase "the undoing heuristic" and gave the new theory the name "Possibility Theory." He then changed it to "Scenario Theory," and then again to "The Theory of Alternative States." In the last notes he made on the subject, he called it "Shadow Theory." "The major point of shadow theory," Amos wrote to himself, "is that the context of alternatives or the possibility set determines our expectations, our interpretations, our recollection and our attribution of reality, as well as the affective states which it induces." Toward the end of his thinking on the subject, he summed up a lot in a single sentence: "Reality is a cloud of possibility, not a point."

It wasn't that Amos had no interest in Danny's thoughts. It was that they were no longer talking in the same room, with the door closed. The conversation that he and Danny were meant to be having together, each was more or less having alone. Because of the new distance between them, each was far more aware where the ideas had come from. "We know who had the idea, because of the physical separation and because the idea is in a letter," Amos would complain to Miles Shore. "Before, we would have picked up the phone at the beginning of an idea. Now you develop an idea and you become committed to them, and they become more personal and you remember you had them. Initially we never had that."

Committed to his new idea, Danny had taken it back rather than let Amos take it apart and remake it into something more like his own. Amos continued to fly to Vancouver every other week, but there was a new tension between them. Amos clearly wished to believe that they might collaborate as they had before. Danny did not. He'd anticipated his own envy and built it into a decision about Amos.

12
THIS CLOUD OF POSSIBILITY

Amos was in Israel on a visit in 1984 when he received the phone call telling him that he'd been given a MacArthur "genius" grant. The award came with two hundred fifty thousand dollars, plus an extra fifty thousand dollars for research, a fancy health care plan, and a press release celebrating Amos as one of the thinkers who had exhibited "extraordinary originality and dedication in their creative pursuits and a marked capacity for self-direction." The only work of Amos's cited in the press release was the work he'd done with Danny. It didn't mention Danny.

Amos disliked prizes. He thought that they exaggerated the differences between people, did more harm than good, and created more misery than joy, as for every winner there were many others who deserved to win, or felt they did. The MacArthur became a case in point. "He wasn't grateful for that prize," said his friend Maya Bar-Hillel, who was with Amos in Jerusalem shortly after the prize was announced. "He was pissed. He said, 'What are these people thinking? How can they give a prize to just one of a winning pair? Do they not realize they are dealing the collaboration a death blow?' " Amos didn't like prizes but he kept on getting them anyway. Before the MacArthur "genius" grant, he had been admitted into the American Academy of Arts and Sciences. Soon after the MacArthur, he received a Guggenheim Fellowship and an invitation to join the National Academy of Sciences. That last honor was seldom bestowed on scientists who weren't U.S. citizens — and it wasn't bestowed upon Danny. There would follow honorary degrees from Yale and the University of Chicago, among others. But the MacArthur was the prize Amos would dwell upon as an example of the damage caused by prizes. "He thought it was myopic beyond forgiveness," said Bar-Hillel. "It was genuine agony. He wasn't putting on a show for me."

Along with the prizes came a steady drizzle of books and articles praising Amos for the work he had done with Danny, as if he had done it alone. When others spoke of their joint work, they put Danny's name second, if they mentioned it at all: Tversky and Kahneman. "You are very generous in giving me credit for articulating the relationship between representativeness and psychoanalysis," Amos wrote to a fellow psychologist who had sent Amos his new journal article. "These ideas, however, were developed in discussions with Danny so you should mention both our names or (if that appears awkward) omit mine." An author of a book credited Amos with noticing the illusory sense of effectiveness felt by Israeli Air Force flight instructors after they'd criticized a pilot. "I am somewhat uncomfortable with the label the 'Tversky effect,' " Amos wrote to the author. "This work has been done in collaboration with my longtime friend and colleague, Daniel Kahneman, so I should not be singled out. In fact, Daniel Kahneman was the one who observed the effect of pilots' training, so if this phenomenon is to be named after a person it should be called the 'Kahneman effect.' "

The American view of his collaboration with Danny mystified Amos. "People saw Amos as the brilliant one and Danny as the careful one," said Amos's friend and Stanford colleague Persi Diaconis. "And Amos would say: 'It's exactly the opposite!' "

Amos's graduate students at Stanford gave

him a nickname: Famous Amos. "You knew that everyone knew him, and you knew everyone wanted to hang out with him," said Brown University professor of psychologist Steven Sloman, who studied with Amos in the late 1980s. The maddening thing is that Amos seemed almost indifferent to the attention. He happily ignored the ever-growing media requests. ("You probably won't be better off after you have appeared on TV than before," he said.) He tossed out as many invitations, unopened, as he acknowledged. None of this arose from a sense of modesty. Amos knew his own value. He didn't need to make a point of not caring what people thought of him; he actually just didn't care all that much. The deal Amos offered the encroaching world was that their interaction was to be on his terms.

And the world accepted the deal. United States congressmen called him for advice on bills they were drafting. The National Basketball Association called to hear his argument about statistical fallacies in basketball. The United States Secret Service flew him to Washington so that he could advise them on how to predict and deter threats to the political leaders under their protection. The North Atlantic Treaty Organization flew him to the French Alps to teach them about how people made decisions in conditions of uncertainty. Amos seemed able to walk into any problem, however alien to him, and make the people dealing with it feel as if he grasped its essence better than they did. The University of Illinois flew him to a conference about metaphorical thinking, for instance, only to have Amos argue that a metaphor was actually a substitute for thinking. "Because metaphors are vivid and memorable, and because they are not readily subjected to critical analysis, they can have considerable impact on human judgment even when they are inappropriate, useless, or misleading," said Amos. "They replace genuine uncertainty about the world with semantic ambiguity. A metaphor is a cover-up."

Danny couldn't help but keep noticing the new attention Amos was receiving for the work they had done together. Economists now wanted Amos at their conferences, but then so did linguists and philosophers and sociologists and computer scientists — even though Amos hadn't the faintest interest in the PC that came with his Stanford office. ("What could I do with computers?" he said, after he'd declined Apple's offer to donate twenty new Macs to the Stanford

Psychology Department.) "You get fed up with not being invited to the same conferences, even when you would not want to go," Danny confessed to the Harvard psychiatrist Miles Shore. "My life would be better if he weren't invited to so many."

In Israel, Danny had been the person real-world people came to when they had some real-world problem. The people in real-world America came to Amos, even when it wasn't obvious that Amos had any reason to know what he was talking about. "He had a hell of an impact on what we did," said Jack Maher, who was in charge of training seven thousand pilots at Delta Air Lines when he sought Amos's help. In the late 1980s, Delta had suffered a series of embarrassing incidents. "We didn't kill anyone," said Maher. "But we'd had people getting lost, people landing at the wrong airports." The incidents nearly always could be traced back to some bad decision made by a Delta captain. "We needed a decision model and I looked for one, but they didn't exist," said Maher. "And Tversky's name kept popping up." Maher met with Amos for a few hours and told him his problems. "He started speaking in math," said Maher. "When he got into linear regression equations I just started to laugh, then he laughed, and stopped doing it." Amos then explained, in plain English, his work with Danny. "He helped us to understand why pilots sometimes made bad decisions," said Maher. "He said, 'You're not going to change people's decision making under duress. You aren't going to stop pilots from making these mental errors. You aren't going to train the decision-making weaknesses out of the pilots.' "

What Delta Air Lines should do, Amos suggested, was to change its decision-making environments. The mental mistakes that led pilots of planes bound for Miami to land boneheadedly in Fort Lauderdale were woven into human nature. People had trouble seeing when their minds were misleading them; on the other hand, they could sometimes see when other people's minds were misleading *them.* But the cockpit culture of a commercial airliner did not encourage people to point out the mental errors of the man in charge. "Captains at the time would be complete autocratic jerks who insisted on running the show," said Maher. The way to stop the captain from landing the plane in the wrong airport, Amos insisted, was to train others in the cockpit to question his judgment. "He changed the way we trained pilots," said

Maher. "We changed the culture in the cockpit and the autocratic jerk became no longer acceptable. Those mistakes haven't happened since."

By the 1980s, the ideas that Danny and Amos had hatched together were infiltrating places the two had never imagined them entering. Success created, among other things, a new market for critics. "We started this unknown field," Amos told Miles Shore in the summer of 1983. "We were shaking trees and challenging the establishment. Now *we* are the establishment. And people are shaking our tree." Those people tended to be self-serious intellectuals. Upon encountering Danny and Amos's work, more than a few academics experienced the sensation that a person feels when a total stranger walks up and begins a sentence, "Don't take this the wrong way, but. . . ." Whatever might follow, you just know that you're not going to like it. The sound of laughter coming from the other side of Amos and Danny's closed door hadn't helped. It caused other intellectuals to wonder about their motives. "The glee is what created the suspicion," said the philosopher Avishai Margalit. "They looked like people standing in front of a monkey cage, making faces at the monkeys. There was too much joy there. They said, 'We're monkeys, too.' But no one believed them. The feeling was that the joy that they have is to trick people. And it stuck. It was a real problem for them."

At a conference back in the early 1970s, Danny was introduced to a prominent philosopher named Max Black and tried to explain to the great man his work with Amos. "I'm not interested in the psychology of stupid people," said Black, and walked away. Danny and Amos didn't think of their work as the psychology of stupid people. Their very first experiments, dramatizing the weakness of people's statistical intuitions, had been conducted on professional statisticians. For every simple problem that fooled undergraduates, they could come up with a more complicated version to fool professors. At least a few professors didn't like the idea of that. "Give people a visual illusion and they say, 'It's only my eyes,' " said Princeton psychologist Eldar Shafir. "Give them a linguistic illusion. They're fooled, but they say, 'No big deal.' Then you give them one of Amos and Danny's examples and they say, 'Now you're insulting me.' "

The first to take their work personally were the psychologists whose work it had trumped. Amos's former teacher Ward Edwards had written the original journal article in 1954 inviting psychologists to investigate the assumptions of economics. Still, he'd never imagined *this* — two Israelis walking into the room and making a mockery of the entire conversation. In late 1970, after reading early drafts of Amos and Danny's papers on human judgment, Edwards wrote to complain. In what would be the first of many agitated letters, he adopted the tone of a wise and indulgent master speaking to his naive pupils. How could Amos and Danny possibly believe that there was anything to learn from putting silly questions to undergraduates? "I think your data collection methods are such that I don't take seriously a single 'experimental' finding you present," wrote Edwards. These students they had turned into their lab rats were "careless and inattentive. And if they are confused and inattentive, they are much less likely to behave more like competent intuitive statisticians." For every supposed limitation of the human mind Danny and Amos had uncovered, Edwards had an explanation. The gambler's fallacy, for instance. If people thought that a coin, after landing on heads five times in a row, was more likely, on the sixth toss, to land on tails, it wasn't because they misunderstood randomness. It was because "people get bored doing the same thing all the time."

Amos took the trouble to answer, almost politely, that first letter from his former professor. "It was certainly a pleasure to read your detailed comments on our papers and to see that, right or wrong, you have not lost any of your old fighting spirit," he began, before describing his former professor as "not cogent." "In particular," Amos continued, "the objections you raised against our experimental method are simply unsupported. In essence, you engage in the practice of criticizing a procedural departure without showing how the departure might account for the results obtained. You do not present either contradictory data or a plausible alternative interpretation of our findings. Instead, you express a strong bias against our method of data collection and in favor of yours. This position is certainly understandable, yet it is hardly convincing."

Edwards was not pleased, but he kept his anger to himself for a few years. "No one wanted to get in a fight with Amos," said the psychologist Irv Biederman. "Not in public! I only once ever saw anyone ever do it. There was this philosopher. At a conference. He gets up to give his talk. He's going to challenge heuristics. Amos was there. When he finished talking Amos got up to rebut. It was like an ISIS beheading. But with humor." Edwards must have sensed, in any open conflict with Amos, the possibility of being on the painful end of an ISIS beheading, with humor. And yet Amos had championed the idea that man was a good intuitive statistician. He needed to say *something.*

In the late 1970s he finally found a principle on which to take a stand: The masses were not equipped to grasp Amos and Danny's message. The subtleties were beyond them. People needed to be protected from misleading themselves into thinking that their minds were less trustworthy than they actually were. "I do not know whether you realize just how far that message has spread, or how devastating its effects have been," Edwards wrote to Amos in September of 1979. "I attended the organizational meeting of the Society for Medical Decision Making one and a half weeks ago. I would estimate that every third paper mentioned your work in passing, mostly as justification for avoiding human intuition, judgment, decision making, and other intellectual processes." Even sophisticated doctors were getting from Danny and Amos only the crude, simplified message that their minds could never be trusted. What would become of medicine? Of intellectual authority? Of experts?

Edwards sent Amos a working draft of his assault on Danny and Amos's work and hoped that Amos would leave him with his dignity. Amos didn't. "The tone is snide, the evaluation of evidence is unfair and there are too many technical difficulties to begin to discuss," Amos wrote, in a curt note to Edwards. "We are in sympathy with your attempt to redress what you regard as a distorted view of man. But we regret that you chose to do so by presenting a distorted view of our work." In his reply, Edwards did a fair impression of a man who has just realized that his fly is unzipped, as he backpedals off a cliff. He offered up his personal problems — they ranged from serious jet lag to "a decade's worth of personal frustrations" — as excuses for his failed paper, and then went on to more or less concede that he wished he'd never written it. "What especially embarrasses me is that after working so long as I did on trying to put this thing together I should have been as blind to its many flaws as I was," he wrote to both Amos and Danny, before saying how he intended to entirely rewrite his paper and hoped very much to avoid any public controversy with them.

Not everyone knew enough to be afraid of Amos. An Oxford philosopher named L. Jonathan Cohen raised a small philosophy-sized ruckus with a series of attacks in books and journals. He found alien the idea that you might learn something about the human mind by putting questions to people. He argued that as man had created the concept of rationality he must, by definition, be rational. "Rational" was whatever most people did. Or, as Danny put it in a letter that he reluctantly sent in response to one of Cohen's articles, "Any error that attracts a sufficient number of votes is not an error at all." Cohen labored to demonstrate that the mistakes discovered by Amos and Danny either were not mistakes or were the result of "mathematical or scientific ignorance" in people, easily remedied by a bit of exposure to college professors. "We both make a living by teaching probability and statistics," Stanford's Persi Diaconis and David Freedman, of the University of California at Berkeley, wrote to the journal *Behavioral and Brain Sciences,* which had published one of Cohen's attacks. "Over and over again we see students and colleagues (and ourselves) making certain kinds of mistakes. Even the same mistake may be repeated by the same person many times. Cohen is wrong in dismissing this as the result of 'mathematical or scientific ignorance.' " But by then it was clear that no matter how often people trained in statistics affirmed the truth of Danny and Amos's work, people who weren't would insist that they knew better.

Upon their arrival in North America, Amos and Danny had published a flurry of papers together. Mostly it was stuff they'd had in the works when they'd left Israel. But in the early 1980s what they wrote together was not done in the same way as before. Amos wrote a piece on loss aversion under both their names, to which Danny added a few stray paragraphs. Danny wrote up on his own what Amos had called "The Undoing Project," titled it "The Simulation Heuristic," and published it with both their names on top, in a book that collected their articles, along with others by students and colleagues. (And then set out to explore the rules of the imagination not with Amos but with his younger colleague at the University of British Columbia, Dale Miller.) Amos wrote an article, addressed directly to economists, to repair technical flaws in prospect theory. "Advances in Prospect Theory," it was called, and though Amos did much of the work on it with his graduate student Rich Gonzalez, it ran as a journal article by Danny and Amos. "Amos said that it had always been Kahneman and Tversky and that this had to be Kahneman and Tversky, and that it would be really strange to add a third person to it," said Gonzalez.

Thus they maintained the illusion that they were still working together, much as before, even as the forces pulling them apart gathered strength. The growing crowd of common enemies failed to unite them. Danny was increasingly uneasy with the attitude Amos took toward their opponents. Amos was built to fight. Danny was built to survive. He shied from conflict. Now that their work was under attack, Danny adopted a new policy: to never review a paper that made him angry. It served as an excuse to ignore any act of hostility. Amos accused Danny of "identifying with the enemy," and he wasn't far off. Danny almost found it easier to imagine himself in his opponent's shoes than in his own. In some strange way Danny contained within himself his own opponent. He didn't need another.

Amos, to be Amos, needed opposition.

Without it he had nothing to triumph over. And Amos, like his homeland, lived in a state of readiness for battle. "Amos didn't have Danny's feeling that we should all think together and work together," said Walter Mischel, who had been the chair of Stanford's Psychology Department when it hired Amos. "He thought, 'Fuck You.' "

That sentiment must have been passing through Amos's mind in the early 1980s even more often than it usually did. The critics publishing attacks on his work with Danny were the least of it. At conferences and in conversations, Amos heard over and over from economists and decision theorists that he and Danny had exaggerated human fallibility. Or that the kinks in the mind that they had observed were artificial. Or only present in the minds of college undergraduates. Or . . . something. A lot of people with whom Amos interacted had big investments in the idea that people were rational. Amos was perplexed by their inability to admit defeat in an argument he had plainly won. "Amos wanted to crush the opposition," said Danny. "It just got under his skin more than it did mine. He wanted to find something to shut people up. Which of course you can never do." Toward the end of 1980, or maybe it was early 1981, Amos came to

Danny with a plan to write an article that would end the discussion. Their opponents might never admit defeat — intellectuals seldom did — but they might at least decide to change the subject. "Winning by embarrassment," Amos called it.

Amos wanted to demonstrate the raw power of the mind's rules of thumb to mislead. He and Danny had stumbled upon some bizarre phenomena back in Israel and never fully explored their implications. Now they did. As ever, they crafted careful vignettes, to reveal the inner workings of the minds of the people they asked to judge them. Amos's favorite was about Linda.

> Linda is 31 years old, single, outspoken and very bright. She majored in philosophy. As a student, she was deeply concerned with issues of discrimination and social justice, and also participated in antinuclear demonstrations.

Linda was designed to be the stereotype of a feminist. Danny and Amos asked: *To what degree does Linda resemble the typical member of each of the following classes?*

1) Linda is a teacher in elementary school.

2) Linda works in a bookstore and takes Yoga classes.
3) Linda is active in the feminist movement.
4) Linda is a psychiatric social worker.
5) Linda is a member of the League of Women voters.
6) Linda is a bank teller.
7) Linda is an insurance salesperson.
8) Linda is a bank teller and is active in the feminist movement.

Danny passed out the Linda vignette to students at the University of British Columbia. In this first experiment, two different groups of students were given four of the eight descriptions and asked to judge the odds that they were true. One of the groups had "Linda is a bank teller" on its list; the other got "Linda is a bank teller and is active in the feminist movement." Those were the only two descriptions that mattered, though of course the students didn't know that. The group given "Linda is a bank teller and is active in the feminist movement" judged it more likely than the group assigned "Linda is a bank teller."

That result was all that Danny and Amos needed to make their big point: The rules of thumb people used to evaluate probability led to misjudgments. "Linda is a bank teller and is active in the feminist movement" could never be more probable than "Linda is a bank teller." "Linda is a bank teller and active in the feminist movement" was just a special case of "Linda is a bank teller." "Linda is a bank teller" included "Linda is a bank teller and activist in the feminist movement" along with "Linda is a bank teller and likes to walk naked through Serbian forests" and all other bank-telling Lindas. One description was entirely contained by the other.

People were blind to logic when it was embedded in a story. Describe a very sick old man and ask people: Which is more probable, that he will die within a week or die within a year? More often than not, they will say, "He'll die within a week." Their mind latches onto a story of imminent death and the story masks the logic of the situation. Amos created a lovely example. He asked people: Which is more likely to happen in the next year, that a thousand Americans will die in a flood, or that an earthquake in California will trigger a massive flood that will drown a thousand Americans? People went with the earthquake.

The force that led human judgment astray in this case was what Danny and Amos had called "representativeness," or the similarity between whatever people were judging and some model they had in their mind of that thing. The minds of the students in the first Linda experiment, latching onto the description of Linda and matching its details to their mental model of "feminist," judged the special case to be more likely than the general one.

Amos wasn't satisfied with stopping there. He wanted to hand the entire list of Lindas to groups of people and have them rank the odds of each line item. He wanted to see if a person who decided that "Linda is a bank teller activist in the feminist movement" also thought it was more probable than "Linda is a bank teller." He wanted to show people making that glaring mistake. "Amos really loved to do that," said Danny. "To win the argument, you want people to actually make mistakes."

Danny was of two minds about this new project, and about Amos. From the moment they had left Israel, they'd been like a pair of swimmers caught in different currents, losing the energy to swim against them. Amos felt the pull of logic, Danny the tug of psychology. Danny wasn't nearly as interested as Amos in demonstrations of human irrationality. His interest in decision theory ended with the psychological insight he brought to it. "There is an underlying debate," said Danny later. "Are we doing psychology or decision theory?" Danny wanted to return to psychology. Plus Danny didn't believe that people would actually make this particular mistake. Seeing the descriptions side by side, they'd realize that it was illogical to say that anyone was more likely to be a bank teller active in the feminist movement than simply a bank teller.

With something of a heavy heart, Danny put what would come to be known as the Linda problem to a class of a dozen students at the University of British Columbia. "Twelve out of twelve fell for it," he said. "I remember I gasped. Then I called Amos from my secretary's phone." They ran many further experiments, with different vignettes, on hundreds of subjects. "We just wanted to look at the boundaries of the phenomenon," said Danny. To explore those boundaries, they finally shoved their subjects' noses right up against logic. They gave subjects the same description of Linda and asked, simply: "Which of the two alternatives is more probable?"

Linda is a bank teller.

Linda is a bank teller and is active in the feminist movement.

Eighty-five percent still insisted that Linda was more likely to be a bank teller in the feminist movement than she was to be a bank teller. The Linda problem resembled a Venn diagram of two circles, but with one of the circles wholly contained by the other. But people didn't see the circles. Danny was actually stunned. "At every step we thought, now *that's* not going to work," he said. And whatever was going on inside people's minds was terrifyingly stubborn. Danny gathered an auditorium full of UBC students and explained their mistake to them. "Do you realize you have violated a fundamental rule of logic?" he asked. "So what!" a young woman shouted from the back of the room. "You just asked for my opinion!"

They put the Linda problem in different ways, to make sure that the students who served as their lab rats weren't misreading its first line as saying "Linda is a bank teller NOT active in the feminist movement." They put it to graduate students with training in logic and statistics. They put it to doctors, in a complicated medical story, in which lay embedded the opportunity to make a fatal error of logic. In overwhelming numbers doctors made the same mistake as undergraduates. "Most participants appeared surprised and dismayed to have made an elementary error of reasoning," wrote Amos and Danny. "Because the conjunction fallacy is easy to expose, people who committed it are left with the feeling that they should have known better."

The paper Amos and Danny set out to write about what they were now calling "the conjunction fallacy" must have felt to Amos like an argument ender — that is, if the argument was about whether the human mind reasoned probabilistically, instead of the ways that Danny and Amos had suggested. They walked the reader through how and why people violated "perhaps the simplest and the most basic qualitative law of probability." They explained that people chose the more detailed description, even though it was less probable, because it was more "representative." They pointed out some places in the real world where this kink in the mind might have serious consequences. Any prediction, for instance, could be made to seem more believable, even as it became less likely, if it was filled with internally consistent details. And any lawyer could at once make a case seem more persuasive, even as he made the truth of it less likely, by adding "representative" details to his description of people and events.

And they showed all over again the power of the mental rules of thumb — these curious forces that they had curiously named "heuristics." To the Linda problem Danny and Amos added another, from work they had done in the early 1970s in Jerusalem.

> In four pages of a novel (about 2,000 words), how many words would you expect to find that have the form _ _ _ _ ing (seven-letter words that end with "ing")? Indicate your best estimate by circling one of the values below:
> 0 1–2 3–4 5–7 8–10 11–15 16+

Then they put to those same people a second question: How many seven-letter words appeared, in that same text, of the form _ _ _ _ _ n _? Of course (of course!) there had to be at least as many seven-letter words with *n* in the sixth position as there were seven-letter words ending in *ing,* as the latter was just one example of the former. People didn't realize that, however. They guessed, on average, that the 2,000-word text contained 13.4 words ending in *ing* and only 4.7 words with *n* in the sixth position. And they did this, Amos and

Danny argued, because it was easier to think of words ending in *ing.* Those words were more available. People's misjudgment of the problem was simply the availability heuristic in action.

The paper was another hit.* "The Linda problem" and "the conjunction fallacy" entered the language. Danny had misgivings, however. The new work was jointly written but it was, he said, "joint and painful." He no longer had the feeling that he and Amos were sharing a mind. Amos had written two entire pages of it by himself in which he sought to define, with greater precision, "representativeness." Danny had wanted to keep the definition vague. Danny was uneasy with the feeling that the paper was less an exploration of a new phenomenon than the forging of a new weapon, to be used by Amos in battle. "It's very Amos," he said. "It's a combative paper. We'll provoke you with this. And we'll show you

* After the article appeared, in the October 1983 issue of *Psychological Review,* the best-selling author and computer scientist Douglas Hofstadter sent Amos his own vignettes. Example: Fido barks and chases cars. Which is Fido more likely to be: (1) a cocker spaniel or (2) an entity in the universe?

that you can't win this argument."

By then their interactions had become fraught. It had taken Danny the longest time to understand his own value. Now he could see that the work Amos had done alone was not as good as the work they had done together. The joint work always attracted more interest and higher praise than anything Amos had done alone. It had apparently attracted this genius award. And yet the public perception of their relationship was now a Venn diagram, two circles, with Danny wholly contained by Amos. The rapid expansion of Amos's circle pushed his borders further and further away from Danny's. Danny felt himself sliding slowly but surely from the small group Amos loved to the large group whose ideas Amos viewed with contempt. "Amos changed," said Danny. "When I gave him an idea he would look for what was good in it. For what was right with it. That, for me, was the happiness in the collaboration. He understood me better than I understood myself. He stopped doing that."

To those close to Amos who glimpsed his interaction with Danny, the wonder wasn't that he and Danny were falling out but that they had ever fallen in. "Danny isn't so easy to access," said Persi Diaconis. "Amos was all out there. The chemistry was so deep, I don't know if it's describable, in a mechanistic way. Each of them was brilliant. And that they did interact, and that they could interact, was a miracle." The miracle did not look as if it was going to survive its removal from the Holy Land.

In 1986 Danny moved with Anne to the University of California at Berkeley — the same university that had told Danny eight years earlier that he was too old for a job. "I really hope the move to Berkeley will open a new era with Danny, with more everyday interaction and less tension," Amos wrote in a letter to a friend. "I am optimistic." When Danny had put himself back on the job market, the year before, he'd discovered that his stock had risen dramatically. He received nineteen offers, including one from Harvard. Anyone who wanted to believe that what ailed Danny was simply an absence of status outside of Israel would have some difficulty explaining what happened next: He went into a depression. "He said he wasn't going to work again," recalled Maya Bar-Hillel, who bumped into Danny soon after he'd moved to Berkeley. "He had no more ideas, everything was getting worse."

Danny's premonition of the ending of a relationship that he had once been unable to imagine ending had a lot to do with the state of his mind. "This is a marriage, a big thing," Danny had said to Miles Shore in the summer of 1983. "We have been working for fifteen years. It would be a disaster to stop. It's like asking people why do they stay married. We would need a strong reason *not* to stay married." But in three short years he'd gone from trying to stay in the marriage to trying to get out. His move to Berkeley had the opposite of the intended effect: Seeing Amos more often only caused him more pain. "We have got to the point that the very thought of telling you of ANY idea that I like (mine or someone else's) makes me anxious," Danny wrote to Amos in March 1987, after one meeting. "An episode such as the one we had yesterday wrecks my life for several days (including anticipation as well as recovery) and I just don't want those anymore. I am not suggesting we stop talking, merely that we show some good sense in adapting to the changes in our relationship."

Amos replied to that letter from Danny with a long letter of his own. "I realize my response style leaves a lot to be desired but you have also become much less interested in objections or criticism, mine or others'," he wrote. "You have become very protective of some ideas and develop an attitude of 'love them or leave them' rather than trying to 'get it right.' One of the things I admired you for most in our joint work was your relentlessness as a critic. You discarded a very attractive treatment of regret (developed mostly by you) because of a single counter-example that hardly anyone (except me) could really appreciate the force of. You prevented us from writing up our work on anchoring because it lacks something etc. I do not see any of this in your attitude to many of your ideas recently." When he'd finished that letter Amos wrote another, to the mathematician Varda Liberman, his friend in Israel. "There is no overlap between the way I see my relationship with Danny and the way he perceives me," he wrote. "What seems to me openness between friends he takes as an insult, and what seems like correct behavior to him is to me unfriendly. It is difficult for him to accept we are different in the eyes of other people."

Danny needed something from Amos. He needed him to correct the perception that they were not equal partners. And he needed it because he suspected Amos shared that perception. "He was too willing to accept a situation that put me in his shadow," said Danny. Amos may have been privately furious that the MacArthur Foundation recognized him and not Danny, but when Danny had called to congratulate him, he had only said offhandedly, "If I hadn't gotten it for this I'd have gotten it for something else." Amos might have written endless recommendations for Danny, and told people privately that he was the greatest living psychologist in the world, but after Danny told Amos that Harvard had asked him to join its faculty, Amos said, "It's me they want." He'd just blurted that out, and then probably regretted saying it — even if he wasn't wrong to think it. Amos couldn't help himself from wounding Danny, and Danny couldn't help himself from feeling wounded. Barbara Tversky occupied the office beside Amos at Stanford. "I would hear their phone calls," she said. "It was worse than a divorce."

The wonder was that Danny didn't simply break off the relationship. By the late 1980s he was behaving like a man caught in some mysterious, invisible trap. Once you had shared a mind with Amos Tversky it was hard to get Amos Tversky out of your mind.

What he did, instead, was get Amos out of his sight, by leaving Berkeley for Princeton, in 1992. "Amos cast a shadow on my life," he said. "I needed to get away. *He possessed my mind.*" Amos couldn't understand this need of Danny's to put three thousand miles between them. He found Danny's behavior mystifying. "Just to give you a small example," Amos wrote to Varda Liberman in early 1994, "there's a book on judgment that came out, and in the introduction there's a passage that says Danny and I are 'inseparable.' This is, of course, an exaggeration. But Danny wrote to the author to tell him that it's an exaggeration and added that 'we haven't had anything to do with each other for a decade.' In the last ten years we published five papers together, and worked on several other projects we never finished (mainly because of me). It's a trivial example but gives you an idea of his state of mind."

For the longest time, even as they went back and forth, the collaboration was over in Danny's mind. And for the longest time, in Amos's mind, it wasn't. "You seem determined to make me an offer that I cannot accept," Danny wrote to Amos in early 1993, after one of Amos's proposals. They remained friends. They found excuses to get together and work through their issues. They kept their troubles so private that most people assumed they were still working together. But Amos liked that fiction better than Danny. He had hopes to write the book that they had agreed to write fifteen years earlier. Danny found ways to let Amos know that wasn't going to happen. "Danny has a new idea how to get the book done," Amos wrote to Liberman in early 1994. "We'll stick together a few papers published recently by each of us, with no connection or structure. This strikes me as so grotesque. It'll look like a collection of work by two people who once worked together and now cannot even coordinate chapters. . . . With the situation as it is I can't find enough positive energy even to start thinking, let alone to write."

If Amos couldn't give Danny what he needed, it was perhaps because he couldn't imagine having the need. The need was subtle. In Israel they'd each had a cucumber. Now Amos had a banana. But the banana wasn't what was provoking Danny to hurl the cucumber in his experimenter's face. Danny didn't need job offers from Harvard or genius awards from the MacArthur Foundation. Those might have helped, but only if they altered Amos's view of him. What Danny needed was for Amos to continue to see him and his ideas uncritically, as he had when they were alone together in a room. If that involved some misperception on Amos's part — some exaggeration of the earthly status of Danny's ideas — well, then, Amos should continue to misperceive. After all, what is a marriage if not an agreement to distort one's perception of another, in relation to everyone else? "I wanted something from *him,* not from the world," said Danny.

In October 1993 Danny and Amos found themselves together at the same conference in Turin, Italy. One evening they went for a walk, and Amos made a request. There was a new critic of their work, a German psychologist named Gerd Gigerenzer, and he was getting a new kind of attention. From the start, those most upset by Danny and Amos's work argued that by focusing only on the mind's errors, they were exaggerating its fallibility. In their talks and writings, Danny and Amos had explained repeatedly that the rules of thumb that the mind used to cope with uncertainty often worked well. But sometimes they didn't; and these specific failures were both interesting in and of themselves and revealing about the mind's inner workings. Why *not* study them? After all, no one complained when you used optical illusions to understand the inner workings of the eye.

Gigerenzer had taken the same angle of attack as most of their other critics. But in Danny and Amos's view he'd ignored the usual rules of intellectual warfare, distorting their work to make them sound even more fatalistic about their fellow man than they were. He also downplayed or ignored most of their evidence, and all of their strongest evidence. He did what critics sometimes do: He described the object of his scorn as he wished it to be rather than as it was. Then he debunked his description. In Europe, Amos told Danny on their walk, Gigerenzer was being praised for "standing up to the Americans," which was odd, as the Americans in this case were Israelis. "Amos says we absolutely must do something about Gigerenzer," recalled Danny. "And I say, 'I don't want to. We'll put in a lot of time. I'll be very angry, and I hate being angry. And it'll be a draw.' And Amos said, 'I've never asked you for anything as a friend. I'm asking you this as a friend.' " And Danny thought: *He's never done that before. I can't really say no.*

It wasn't long before he wished that he had. Amos didn't merely want to counter Gigerenzer; he wanted to destroy him.

("Amos couldn't mention Gigerenzer's name without using the word 'sleazeball,' " said UCLA professor Craig Fox, Amos's former student.) Danny, being Danny, looked for the good in Gigerenzer's writings. He found this harder than usual to do. He'd avoided even visiting Germany until the 1970s. When he finally visited, he traveled the streets entertaining a strange and vivid fantasy that the houses were all empty. But Danny didn't like being angry at people, and he contrived not to feel anger toward their new German critic. He even found himself in some slight sympathy with Gigerenzer on one point: the Linda problem. Gigerenzer had shown that, by changing the simplest version of the problem, he could lead people to the correct answer. Instead of asking people to rank the likelihood of the two descriptions of Linda, he asked: *To how many of 100 people who are like Linda do the following statements apply?* When you gave people that hint, they realized that Linda was more likely to be a bank teller than a bank teller active in the feminist movement. But then Danny and Amos already knew that. They'd written as much, with less emphasis, in their original paper.

At any rate, they'd always thought that the most outrageous version of the Linda problem was superfluous to the point they were making — that people judged by representativeness. The very first experiment, like their earlier work on human judgment, showed that plainly enough, and yet Gigerenzer didn't mention it. He had found their weakest evidence and attacked it, as if it were the only evidence they had. Combining his peculiar handling of the evidence with what struck Danny and Amos as a willful misreading of their words, Gigerenzer gave talks and wrote articles with provocative titles like "How to Make Cognitive Illusions Disappear." "Making cognitive illusions disappear was really making us disappear," said Danny. "He was obsessed. I've never seen anything like it."

Gigerenzer came to be identified with a strain of thought known as evolutionary psychology, which had in it the notion that the human mind, having adapted to its environment, must be very well suited to it. It certainly wouldn't be susceptible to systematic biases. Amos found that notion absurd. The mind was more like a coping mechanism than it was a perfectly designed tool. "The brain appears to be programmed, loosely speaking, to provide as much certainty as it can," he once said, in a talk to a group of Wall Street executives. "It is apparently designed to make the best possible case for a given interpretation rather than to represent all the uncertainty about a given situation." The mind, when it dealt with uncertain situations, was like a Swiss Army knife. It was a good enough tool for most jobs required of it, but not exactly suited to anything — and certainly not fully "evolved." "Listen to evolutionary psychologists long enough," Amos said, "and you'll stop believing in evolution."

Danny wanted to understand Gigerenzer better, perhaps even reach out to him. "I was always more sympathetic than Amos to the critics," said Danny. "I tend to almost automatically take the other side." He wrote to Amos to say that he thought the man might be in the grip of some mind-warping emotion. Perhaps they should sit down together and see if they might lead him to reason. "Even if it were true you should not say it," Amos shot back, "and I doubt that it is true. An alternative hypothesis to which I lean is that he is much less emotional than you think, and that he is acting like a lawyer trying to score points to impress the uninformed jury, with little concern for the truth. . . . This does not make me like him more but it makes his behavior easier to understand."

Danny agreed to help Amos "as a friend," but it wasn't long before Amos, once again, was making him miserable. They wrote, and rewrote, drafts of a response to Gigerenzer but at the same time wrote and rewrote the dispute between each other. Danny's language was always too soft for Amos, and Amos's language too harsh for Danny. Danny was always the appeaser, Amos the bully. They could agree on seemingly nothing. "I am so desperately unhappy at the idea of revisiting the GG postscript that I am almost ready to have a chance device (or a set of three judges) decide between our two versions," Danny wrote to Amos. "I don't feel like arguing about it, and what you write feels alien to me." Four days later, after Amos had pressed on, Danny added, "On a day on which they announce the discovery of 40 billion new galaxies we argue about six words in a postscript. . . . It is remarkable how ineffective the number of galaxies is as an argument for giving up in the debate between 'repeat' and 'reiterate.' " And then: "Email is the medium of choice at this stage. Every conversation leaves me upset for a long time, which I cannot afford." To which Amos replied, "I do not get your sensitivity metric. In general, you are the most open minded and least defensive person I know. At the same time, you can get really upset because I rewrote a paragraph you like, or because you chose to interpret a totally harmless comment in an unintended negative way."

One night in New York, while staying in an apartment with Amos, Danny had a dream. "And in this dream the doctor tells me I have six months to live," he recalled. "And I said, 'This is wonderful because no one would expect me to spend the last six months of my life working on this garbage.' The next morning I told Amos." Amos looked at Danny and said, "Other people might be impressed but I am not." *Even if you had only six months to live I'd expect you to finish this with me.* Not long after that exchange, Danny read a list of the new members of the National Academy of Sciences, to which Amos had belonged for nearly a decade. Once again, Danny's name wasn't on the list. Once again, the differences between them were there for all to see. "I asked him, why haven't you put me forward?" said Danny. "But I knew why." Had their situations been reversed, Amos would never have wanted to be given anything on the strength of his friendship with Danny. At bottom, Amos saw Danny's need as weakness. "I said, 'That's not how friends behave,' " said Danny.

And with that Danny left. Walked out. Never mind Gerd Gigerenzer, or the collaboration. He told Amos that they were no longer even friends. "I sort of divorced him," said Danny.

Three days later Amos called Danny. He'd just received some news. A growth that doctors had discovered in his eye had just been diagnosed as malignant melanoma. The doctors had scanned his body and found it riddled with cancer. They were now giving him, at best, six months to live. Danny was the second person he'd called with the news. Hearing that, something inside Danny gave. "He was saying, 'We're friends, whatever you think we are.' "

CODA: BORA-BORA

Consider the following scenario.

> Jason K. is a fourteen-year-old homeless boy who lives in a large American city. He is shy and withdrawn but extremely resourceful. His father was murdered when he was young; his mother is an addict. Jason takes care of himself, sleeping sometimes on the sofas at friends' apartments but mostly on the streets. He manages to stay in school until the ninth grade. He often goes hungry. One day in 2010 he accepts an offer from a local gang to sell drugs, and drops out of school. A few weeks later, the night before his fifteenth birthday, he is shot and killed. He was unarmed when he died.

We are seeking ways to "undo" Jason K.'s death. Rank the following in order of their likelihood.

1) Jason's father was not murdered.
2) Jason carried a gun and was able to protect himself.
3) The U.S. federal government made it easier for homeless kids to obtain the free breakfast and lunch to which they are entitled. Jason never went hungry, and remained in school.
4) A lawyer steeped in the writings of Amos Tversky and Daniel Kahneman took a federal government job in 2009. Drawing upon Kahneman and Tversky's work, he pushed for changes in the rules, so that homeless kids no longer needed to enroll in the school meal program. Instead they automatically received free breakfast and lunch. Jason never went hungry, and remained in school.

If you found #4 more probable than #3, you violated perhaps the simplest and most fundamental law of probability. But you're also onto something. The lawyer's name is Cass Sunstein.

Among its other consequences, the work that Amos and Danny did together awakened economists and policy makers to the importance of psychology. "I became a believer," said Nobel Prize–winning economist Peter Diamond of Danny and Amos's work. "It's all true. This stuff is not just lab stuff. It's capturing reality, and it's important to economists. And I spent years thinking of how to use it — and failing." By the early 1990s a lot of people thought it was a good idea to bring together psychologists and economists, to allow them to get to know each other better. But as it turned out, they didn't particularly want to know each other better. Economists were brash and self-assured. Psychologists were nuanced and doubtful. "Psychologists as a rule will only interrupt a presentation for clarification," says psychologist Dan Gilbert. "Economists will interrupt to show how smart they are." "In economics it is completely normal to be rude," says economist George Loewenstein. "We tried to create a psychology and economics seminar at Yale. We had our first meeting. The psychologists came out completely bruised. We never had a second meeting." In the early 1990s, Amos's former student Steven Sloman invited an equal number of economists and psychologists to a conference in France. "And I swear to God I spent three-quarters of my time telling the economists to shut up," said Sloman. "The problem," says Harvard social psychologist Amy Cuddy, "is that psychologists think economists are immoral and economists think psychologists are stupid."

In the academic culture war triggered by Danny and Amos's work, Amos served as a strategic advisor. At least some of his sympathies were with the economists. Amos's mind had always clashed with most of psychology. He didn't like emotion, as a subject. His interest in the unconscious mind was limited to a desire to prove it didn't exist. He was like a man in stripes wandering a land settled by people dressed in plaids and polka dots. Like the economists, he preferred neat formal models to mixed-chocolate boxes of psychological phenomena. Like them, he found it completely normal to be rude. And, like them, he had worldly ambitions for his ideas. Economists sought influence in the arenas of finance and business and public policy. Psychologists hardly ever entered those arenas. That was about to change.

Danny and Amos both saw that there was no point trying to infiltrate economics from psychology. The economists would just ignore intruders. What were needed were young economists with an interest in psychology. Almost magically, after Amos and Danny arrived in North America, they began to appear. George Loewenstein was a good example. A trained economist disillusioned by the psychological sterility of economic models, Loewenstein read Amos and Danny's work and thought: Wait, maybe I want to be a psychologist! As he happened to be Sigmund Freud's great-grandson, this was an even more complicated than usual thought. "I had tried to escape the family's past," said Loewenstein. "I realized I had never taken a single class in what really interested me." He approached Amos and asked him for advice: Should he move from economics to psychology? "Amos said, 'You should stay in economics — we need you there.' He already knew in 1982 that he was starting a movement. And he needed people inside economics."

The argument that Danny and Amos started would spill over into law and public policy. Psychology would use economics to enter these places and others. Richard Thaler — the first frustrated economist to stumble onto Danny and Amos's work and pursue its consequences for economics single-mindedly — would help to create a new field, and give it the name "behavioral economics." "Prospect Theory," scarcely cited in the first decade after its publication, would become, by 2010, the second most cited paper in all of economics. "People tried to ignore it," said Thaler. "Old economists never change their minds." By 2016 every tenth paper published in economics would have a behavioral angle to it, which is to say it had at least a whisper of the work of Danny and Amos. And Richard Thaler would have just stepped down from his tenure as president of the American Economic Association.

Cass Sunstein had been a young law professor at the University of Chicago when he came upon Thaler's first war cry on psychology's behalf. A paper Thaler had titled in his mind "Stupid Shit That People Do" he'd finally published as "Toward a Positive Theory of Consumer Choice." Thaler's bibliography led Sunstein directly to the article written by Danny and Amos in *Science* about judgment, and to "Prospect Theory." "For a lawyer both of these were difficult," said Sunstein. "I had to read them more than once. But I remember the feeling: It was an explosion of lightbulbs. You have thoughts in your mind and you read something that immediately puts them in order and it's electrifying." In 2009, at the invitation of President Obama, Sunstein went to work at the White House. There he oversaw the Office of Information and Regulatory Affairs and made scores of small changes that had big effects on the daily lives of all Americans.

The changes Sunstein made had a unifying theme: They sprang directly or indirectly from the work of Danny and Amos. You couldn't say that Danny and Amos's work led President Obama to ban federal employees from texting while driving, but it wasn't hard to draw a line from their work to that act. The federal government now became sensitive to both loss aversion and framing effects: People didn't choose between things, they chose between descriptions of things. The fuel labels on new automobiles went from listing only miles per gallon to including the number of gallons a car consumed every hundred miles. What used to be called the food pyramid became MyPlate, a graphic of a dinner plate with divisions for each of the five food groups, and it was suddenly easier for Americans to see what made for a healthy diet. And so on. Sunstein argued that the government needed, alongside its Council of Economic Advisers, a Council of Psychological Advisers. He wasn't alone. By the time Sunstein left the White House, in

2015, calls for a greater role for psychologists, or at any rate for psychological insight, were coming from inside governments across the world.

Sunstein was particularly interested in what was now being called "choice architecture." The decisions people made were driven by the way they were presented. People didn't simply know what they wanted; they took cues from their environment. They *constructed* their preferences. And they followed paths of least resistance, even when they paid a heavy price for it. Millions of U.S. corporate and government employees had woken up one day during the 2000s and found they no longer needed to enroll themselves in retirement plans but instead were automatically enrolled. They probably never noticed the change. But that alone caused the participation in retirement plans to rise by roughly 30 percentage points. Such was the power of choice architecture. One tweak to the society's choice architecture made by Sunstein, once he'd gone to work in the U.S. government, was to smooth the path between homeless children and free school meals. In the school year after he left the White House, about 40 percent more poor kids ate free school lunches than had done so before, back when they or some adult acting on their behalf had to take action and make choices to get them.

Even in Canada, Don Redelmeier still heard the sound of Amos in his head. He'd been back from Stanford for several years, but Amos's voice was so clear and overpowering that it made it hard for Redelmeier to hear his own. Redelmeier could not pinpoint the precise moment that he sensed that his work with Amos was not all Amos's doing — that it had some Redelmeier in it, too. His sense of his own distinct value began with a simple question — about homeless people. The homeless were a notorious drag on the local health care system. They turned up in emergency rooms more often than they needed to. They were a drain on resources. Every nurse in Toronto knew this: If you see a homeless person wander in, hustle him out the door as fast as you can. Redelmeier wondered about the wisdom of that strategy.

And so, in 1991, he created an experiment. He arranged for large numbers of college students who wanted to become doctors to be given hospital greens and a place to sleep near the emergency room. Their job was to serve as concierges to the homeless. When a homeless person entered the emergency room, they were to tend to his every need. Fetch him juice and a sandwich, sit down and talk to him, help arrange for his medical care. The college students worked for free. They loved it: They got to pretend to be doctors. But they serviced only half of the homeless people who entered the hospital. The other half received the usual curt and dismissive service from the nursing staff. Redelmeier then tracked the subsequent use of the Toronto health care system by all the homeless people who had visited his hospital. Unsurprisingly, the group that received the gold-plated concierge service tended to return slightly more often to the hospital where they had received it than the unlucky group. The surprise was that their use of the greater Toronto health care system declined. When homeless people felt taken care of by a hospital, they didn't look for other hospitals that might take care of them. The homeless said, "That was the best that can be done for me." The entire Toronto health care system had been paying a price for its attitude to the homeless.

A part of good science is to see what everyone else can see but think what no one else has ever said. Amos had said that to him, and it had stuck in Redelmeier's mind. By the mid-1990s, in startling ways, Redelmeier was taking what everyone could see and thinking to say what no one had ever said. For instance, one day he had a phone call from an AIDS patient who was suffering side effects of medication. In the middle of the conversation, the patient cut him off and said, "I'm sorry, Dr. Redelmeier, I have to go. I just had an accident." The guy had been talking to him on a cell phone while driving. Redelmeier wondered: Did talking on a cell phone while driving increase the risk of accident?

In 1993, he and Cornell statistician Robert Tibshirani created a complicated study to answer the question. The paper they wrote, in 1997, proved that talking on a cell phone while driving was as dangerous as driving with a blood alcohol level at the legal limit. A driver talking on a cell phone was four times as likely as a driver who wasn't to be involved in a crash, *whether or not he held the phone in his hands.* That paper — the first to establish, rigorously, the connection between cell phones and car accidents — spurred calls for regulation around the world. It would take another, even more complicated study to determine just how many thousands of lives it saved.

The study also piqued Redelmeier's interest in what happened inside the mind of a person behind the wheel of a car. The doctors in the Sunnybrook trauma center assumed that their jobs began when the human beings mangled on nearby Highway 401 arrived in the emergency room. Redelmeier thought it was insane for medicine not to attack the problem at its source. One point two million people on the planet died every year in car accidents, and multiples of that were maimed for life. "One point two million deaths a year worldwide," said Redelmeier. "One Japanese tsunami every day. Pretty impressive for a cause of death that was unheard-of one hundred years ago." When exercised behind the wheel of a car, human judgment had irreparable consequences: That idea now fascinated Redelmeier. The brain is limited. There are gaps in our attention. The mind contrives to make those gaps invisible to us. We think we know things we don't. We think we are safe when we are not. "For Amos it was one of the core lessons," said Redelmeier. "It's not that people think they are perfect. No, no: They can make mistakes. It's that they don't appreciate the extent to which they are fallible. *'I've had three or four drinks. I might be 5 percent off my game.'* No! You are actually 30 percent off your game. This is the mismatch that leads to ten thousand fatal accidents in the United States every year."

It is sometimes easier to make the world a better place than to prove you have made the world a better place.

Amos had said that, too. "Amos gave everyone permission to accept human error," said Redelmeier. That was how Amos made the world a better place, though it was impossible to prove. The spirit of Amos was now present in everything Redelmeier did. It was present in his article about the dangers of driving while speaking on a cell phone — which Amos had read and commented upon. That was the article Redelmeier was working on when the call came with the news that Amos had died.

Amos told very few people that he was dying, and, to those he did tell, he gave instructions not to spend a lot of time talking to him about it. He received the news in February 1996. From then on he spoke of his life in the past tense. "He called me when the doctor told him that it was the end of it," said Avishai Margalit. "I came to see him. And he fetched me from the airport. And we were on our way to Palo Alto. And we stopped somewhere on the road, with a view, and talked, about death and about life. It was important to him that he had his death under control. And the feeling was that he was talking not about himself. Not about *his* death. There was a kind of stoic distance that was astonishing. He said, 'Life is a book. The fact that it was a short book doesn't mean it wasn't a good book. It was a very good book.' " Amos seemed to understand that an early death was the price of being a Spartan.

In May Amos gave his final lecture at Stanford, about the many statistical fallacies in professional basketball. His former graduate student and collaborator Craig Fox asked Amos if he would like for it to be videotaped. "He thought about it and said, 'No, I don't think so,' " recalled Fox. With one exception, Amos didn't change his routine, or even his interactions with those around him, in any way. The exception was that, for the first time, he spoke of his experience of war. For instance, he told Varda Liberman the story of how he had saved the life of the soldier who had fainted on top of the bangalore mine. "He said this one event in a way kind of shaped his entire life," said Liberman. "He said, 'Once I did that, I felt obliged to keep this image of hero. I did that, now I have to live up to it.' "

Most people with whom Amos interacted never even suspected he was ill. To a graduate student who asked if he would supervise his dissertation, Amos simply said, "I'm going to be very busy the next few years," and sent him on his way. A few weeks before his death, he called his old friend Yeshu Kolodny in Israel. "He was very impatient, which never happened," recalled Kolodny. "He said, 'Listen, Yeshu, I'm dying. I take it not tragically. But I don't want to talk to anyone. I need you to call our friends and tell them — and tell them not to call or visit.' " To his rule against visitors Amos made an exception for Varda Liberman, with whom he was finishing a textbook. He made another for Stanford president Gerhard Casper — but only because he'd gotten wind of Stanford's plan to commemorate him, with a lecture series or a conference in his name. "Amos told Casper, 'You can do anything you want,' " recalled Liberman. " 'But I beg you, don't have a conference in my name with mediocre people who will talk about their work and how it is "related" to mine. Just put my name on a building. Or a room. Or a bench. You can put it on anything that is not moving.' "

He accepted very few phone calls. One he

took came from the economist Peter Diamond. "I learned he was dying," said Diamond. "And I learned he wasn't taking phone calls. But I had just finished my report to the Nobel Committee." Diamond wanted to let Amos know that he was on a very short list for the Nobel Prize in economics, to be awarded in the fall. But the Nobel Prize was awarded only to the living. He didn't recall what Amos said to that, but Varda Liberman was in the room when Amos took the call. "I thank you very much for letting me know," she heard Amos say. "I can assure you that the Nobel Prize is not on the list of things I'm going to miss."

He spent the last week of his life at home, with his wife and children. He'd obtained the drugs he needed to end his own life, when he felt it was no longer worth living, and found ways to let his children know what he planned to do, without coming out and saying it. ("What do you think of euthanasia?" he asked his son Tal casually.) Toward the end, his mouth turned blue; his body was bloated. He never took painkillers. On May 29, Israel held an election for prime minister, and a militaristic Benjamin Netanyahu defeated Shimon Peres. "So I won't see peace in my lifetime," Amos said, upon hearing the news. "But I was never going to see peace in my lifetime." Late on the night of June 1, his children heard from their father's bedroom the sound of footsteps and his voice. Talking, perhaps to himself. Thinking. On the morning of June 2, 1996, Amos's son Oren entered his father's bedroom and found him dead.

His funeral was a blur. It had an unreal feeling to it. The people in attendance could imagine many things, but they had trouble imagining Amos Tversky dead. "Death is unrepresentative of Amos," said his friend Paul Slovic. Amos's Stanford colleagues, who had come to think of Danny as a figure from the distant past, were shocked when he appeared and approached the front of the synagogue. ("It was like seeing a fucking ghost," said one.) "He seemed disoriented, almost shell-shocked," recalled Avishai Margalit. "There was a feeling of unfinished business." In a room filled with people dressed in dark suits, Danny had arrived in shirtsleeves, as he would have done for an Israeli funeral. That struck people as odd: He didn't seem to know where he was. But no one thought it was anything but correct that Danny delivered the main eulogy. "It was clear that he was the one to talk," said Margalit.

■ ■ ■ ■

Their final conversations had been mostly about their work. But not all of them. There were things Amos wanted to say to Danny. He wanted to tell him that no one had caused him more pain in his life. To stop himself from echoing the sentiment, Danny had to bite his tongue. He also said that Danny was, even now, the person he most wanted to talk to. "He said I'm the one he's most comfortable talking to, because I'm not afraid of death," recalled Danny. "He knew I'm ready to die anytime."

As Amos approached his death, Danny spoke to him nearly every day. He wondered aloud at Amos's desire to keep on living exactly as he had, and his disinterest in fresh, new experiences. "What am I going to do, go to Bora-Bora?" Amos had replied. From that moment Danny lost any interest he might have had in ever visiting Bora-Bora. The mention of the name would forever trigger an uneasy ripple in his mind. After Amos had told him that he was dying, Danny had suggested that they write something together — an introduction to a collection of their old papers. Amos had died before they could finish. In their final conversation, Danny told Amos that he dreaded the thought of writing something under Amos's name of which Amos might disapprove. "I said, 'I don't trust what I'm going to do,' " Danny said. "And he said, 'You will just have to trust in the model of me that is in your mind.' "

Danny remained at Princeton, where he had gone to escape Amos. After Amos's death, Danny's phone rang more often than ever. Amos might be gone, but their work lived on, and it was getting more and more attention. And when people spoke of it they no longer said "Tversky and Kahneman." People began to refer to "Kahneman and Tversky." Then, in the fall of 2001, Danny received an invitation to visit Stockholm and speak at a conference. Members of the Nobel Committee would attend, along with leading economists. All the speakers but Danny were economists. Like Danny, they were all pretty obviously under consideration for the prize. "It was an audition," said Danny. He worked hard to prepare his talk, which he knew had to be about something other than the work he had done with Amos. Some of his friends found that odd, as it was the joint work with Amos that had caught the interest of the Nobel Committee. "I was invited for the joint work," said

Danny, "but I needed to show that I on my own am good enough. The question wasn't, was the work worthy? The question was, am *I* worthy?"

Danny didn't usually prepare his talks. He'd once given a college commencement speech entirely off-the-cuff, and no one seemed to realize that he hadn't thought about what he was going to say until he sat on the dais waiting to be announced. That talk in Stockholm he'd really worked on. "I sweated it out to such an extent that I spent a lot of time choosing the exact color of the background for the slides," he said. His subject was happiness. He spoke of the ideas that he most regretted not exploring together with Amos. How people's anticipation of happiness differed from the happiness they experienced, and how both differed from the happiness they remembered. How you could measure these things — by, say, questioning people before, during, and after painful colonoscopies. If happiness was so malleable, it made a mockery of economic models that were premised on the idea that people maximized their "utility." What, exactly, was to be maximized?

After his talk, Danny returned to Princeton. He had the idea that, if he was ever to be given a Nobel Prize, it would be the following year. They'd seen and heard him in the flesh. They'd judge him worthy or not.

All potential winners were aware of the day the call from Stockholm would come, in the early morning, were it to come at all. On October 9, 2002, Danny and Anne sat in their home in Princeton, both waiting and not waiting. Danny was actually writing a reference for one of his star graduate students, Terry Odean. He honestly hadn't thought much about what he would do if he won a Nobel Prize. Or, rather, he had specifically not allowed himself to think much about what he would do if he won a Nobel Prize. As a child during the war, he'd cultivated an active fantasy life. He would play out elaborate scenes with himself at the center of them. He imagined himself single-handedly winning the war and ending it, for example. But because he was Danny, he made a rule about his fantasy life: He never fantasized about something that might happen. He established this private rule for his imagination once he realized that, after he had fantasized about something that might actually happen, he lost his drive to make it happen. His fantasies were so vivid that "it was as if you actually had it," and if you actually had it, why would you bother to work hard to get it?

He'd never end the war that killed his father, so what did it matter if he created an elaborate scenario in which he won it single-handedly?

Danny had not allowed himself to imagine what he would do if he were ever given a Nobel Prize. Which was just as well, as the phone didn't ring. At some point Anne got up and said, a bit sadly, "Oh well." Every year there were disappointed people. Every year there were old people waiting by phones. Anne went off to exercise and left Danny alone. He'd always been good at preparing himself for not getting what he wanted, and in the grand scheme of things this was not a hard blow. He was fine with who he was and what he had done. He could now safely imagine what he would have done had he won the Nobel Prize. He would have brought Amos's wife and children with him. He would have appended to his Nobel lecture his eulogy of Amos. He would have carried Amos to Stockholm with him. He would have done for Amos what Amos could never do for him. There were many things Danny would have done, but now he had things to do. He went back to writing his enthusiastic reference for Terry Odean.

Then the phone rang.

A NOTE ON SOURCES

Papers written for social science journals are not intended for public consumption. For a start, they're instinctively defensive. The readers of academic papers, in the mind's eye of their authors, are at best skeptical, and more commonly hostile. The writers of these papers aren't trying to engage their readers, much less give them pleasure. They're trying to *survive* them. As a result, I found that I was able to get a clearer, more direct, and more enjoyable understanding of the ideas in academic papers by speaking with their authors than by reading the papers themselves — though of course I read the papers, too.

The academic papcrs of Tversky and Kahneman are an important exception. Even as they wrote for a narrow academic audience, Danny and Amos seemed to sense a general reader waiting for them, in the future. Danny's book *Thinking, Fast and Slow* was openly directed at the general reader, and that helped this general reader in many ways. Actually, I watched Danny agonize over his book for several years, and even read early drafts of some of it. Everything Danny wrote, like everything he said, was full of interest. Still, every few months he'd be consumed with despair, and announce that he was giving up writing altogether — before he destroyed his own reputation. To forestall his book's publication he *paid* a friend to find people who might convince him not to publish it. After its publication, when it landed on the *New York Times* bestseller list, he bumped into another friend, who later described what must be the oddest response any author has ever had to his own success. "You'll never believe what happened," said Danny incredulously. "Those people at the *New York Times* made a mistake and put my book on the bestseller list!" A few weeks later, he bumped into the same friend. "It's unbelievable what is going on," said Danny. "Because those people at the *New York Times* made that mistake and put my book on their bestseller list, they've had to keep it there!"

I would encourage anyone interested in my book to read Danny's book, too. For those whose thirst for psychology remains unquenched, I'd recommend two other books, which helped me come to grips with the field. The eight-volume *Encyclopedia of Psychology* will answer just about any question you might have about psychology, clearly and directly. The nine-volume (and counting) *A History of Psychology in Autobiography* will answer just about any question you might have about psychologists, though less directly. The first volume of this remarkable series was published in 1930, and it continues to motor along, fueled by an endlessly renewable source of energy: the need felt by psychologists to explain why they are the way they are.

At any rate, in grappling with my subject, I obviously leaned on the work of others. Here are those I leaned on:

Introduction: The Problem That Never Goes Away

Thaler, Richard H., and Cass R. Sunstein. "Who's on First." *New Republic,* August 31, 2003. https://newrepublic.com/article/61123/whos-first.

Chapter 1: Man Boobs

Rutenberg, Jim. "The Republican Horse Race Is Over, and Journalism Lost." *New York Times,* May 9, 2016.

Chapter 2: The Outsider

Meehl, Paul E. *Clinical versus Statistical Prediction.* Minneapolis: University of Minnesota Press, 1954.

———. "Psychology: Does Our Heterogeneous Subject Matter Have Any Unity?" *Minnesota Psychologist* 35 (1986): 3–9.

Chapter 3: The Insider

Edwards, Ward. "The Theory of Decision Making." *Psychological Bulletin* 51, no. 4 (1954): 380–417. http://worthylab.tamu.edu/courses_files/01_edwards_1954.pdf.

Guttman, Louis. "What Is Not What in Statistics." *Journal of the Royal Statistical Society* 26, no. 2 (1977): 81–107. http://www.jstor.org/stable/2987957.

May, Kenneth. "A Set of Independent Necessary and Sufficient Conditions for Simple Majority Decision." *Econometrica* 20, no. 4 (1952): 680–84.

Rosch, Eleanor, Carolyn B. Mervis,

Wayne D. Gray, David M. Johnson, and Penny Boyes-Braem. "Basic Objects in Natural Categories." *Cognitive Psychology* 8 (1976): 382–439. http://www.cns.nyu.edu/~msl/courses/2223/Readings/Rosch-CogPsych1976.pdf.

Tvcrsky, Amos. "The Intransitivity of Preferences." *Psychological Review* 76 (1969): 31–48.

———. "Features of Similarity." *Psychological Review* 84, no. 4 (1977): 327–52. http://www.ai.mit.edu/projects/dm/Tversky-features.pdf.

Chapter 4: Errors

Hess, Eckhard H. "Attitude and Pupil Size." *Scientific American,* April 1965, 46–54.

Miller, George A. "The Magical Number Seven, Plus or Minus Two: Some Limits on Our Capacity for Processing Information." *Psychological Review* 63 (1956): 81–97.

Chapter 5: The Collision

Friedman, Milton. "The Methodology of Positive Economics." In *Essays in Positive Economics,* edited by Milton Friedman, 3–46. Chicago: University of Chicago

Press, 1953.
Krantz, David H., R. Duncan Luce, Patrick Suppes, and Amos Tversky. *Foundations of Measurement* — Vol. I: *Additive and Polynomial Representations;* Vol. II: *Geometrical, Threshold, and Probabilistic Representations;* Vol III: *Representation, Axiomatization, and Invariance.* San Diego and London: Academic Press, 1971–90; repr., Mineola, NY: Dover, 2007.
Tversky, Amos, and Daniel Kahneman. "Belief in the Law of Small Numbers." *Psychological Bulletin* 76, no. 2 (1971): 105–10.

Chapter 6: The Mind's Rules

Glanz, James, and Eric Lipton. "The Height of Ambition." *New York Times Magazine,* September 8, 2002.
Goldberg, Lewis R. "Simple Models or Simple Processes? Some Research on Clinical Judgments." *American Psychologist* 23, no. 7 (1968): 483–96.
———. "Man versus Model of Man: A Rationale, Plus Some Evidence, for a Method of Improving on Clinical Inferences." *Psychological Bulletin* 73, no. 6 (1970): 422–32.

Hoffman, Paul J. "The Paramorphic Representation of Clinical Judgment." *Psychological Bulletin* 57, no. 2 (1960): 116–31.
Kahneman, Daniel, and Amos Tversky. "Subjective Probability: A Judgment of Representativeness." *Cognitive Psychology* 3 (1972): 430–54.
Meehl, Paul E. "Causes and Effects of My Disturbing Little Book." *Journal of Personality Assessment* 50, no. 3 (1986): 370–75.
Tversky, Amos, and Daniel Kahneman. "Availability: A Heuristic for Judging Frequency and Probability." *Cognitive Psychology* 5, no. 2 (1973): 207–32.

Chapter 7: The Rules of Prediction

Fischhoff, Baruch. "An Early History of Hindsight Research." *Social Cognition* 25, no. 1 (2007): 10–13.
Howard, R. A., J. E. Matheson, and D. W. North. "The Decision to Seed Hurricanes." *Science* 176 (1972): 1191–1202. http://www.warnernorth.net/hurricanes.pdf.
Kahneman, Daniel, and Amos Tversky. "On the Psychology of Prediction." *Psychological Review* 80, no. 4 (1973): 237–51.

Meehl, Paul E. "Why I Do Not Attend Case Conferences." In *Psychodiagnosis: Selected Papers,* edited by Paul E. Meehl, 225–302. Minneapolis: University of Minnesota Press, 1973.

Chapter 8: Going Viral

Redelmeier, Donald A., Joel Katz, and Daniel Kahneman. "Memories of Colonoscopy: A Randomized Trial." *Pain* 104, nos. 1–2 (2003): 187–94.

Redelmeier, Donald A., and Amos Tversky. "Discrepancy between Medical Decisions for Individual Patients and for Groups." *New England Journal of Medicine* 322 (1990): 1162–64.

———. Letter to the editor. *New England Journal of Medicine* 323(1990): 923. http://www.nejm.org/doi/pdf/10.1056/NEJM199009273231320.

———. "On the Belief That Arthritis Pain Is Related to the Weather." *Proceedings of the National Academy of Sciences* 93, no. 7(1996): 2895–96. http://www.pnas.org/content/93/7/2895.full.pdf.

Tversky, Amos, and Daniel Kahneman. "Judgment under Uncertainty: Heuristics and Biases." *Science* 185 (1974): 1124–31.

Chapter 9: Birth of the Warrior Psychologist

Allais, Maurice. "Le Comportement de l'homme rationnel devant le risque: critique des postulats et axiomes de l'école américaine." *Econometrica* 21, no. 4 (1953): 503–46. English summary: https://goo.gl/cUvOVb.

Bernoulli, Daniel. "Specimen Theoriae Novae de Mensura Sortis." *Commentarii Academiae Scientiarum Imperialis Petropolitanae, Tomus V* [Papers of the Imperial Academy of Sciences in Petersburg, Vol. V], 1738, 175–92. Dr. Louise Sommer of American University did apparently the first translation into English: for *Econometrica* 22, no. 1 (1954): 23–36. See also Savage (1954) and Coombs, Dawes, and Tversky (1970).

Coombs, Clyde H., Robyn M. Dawes, and Amos Tversky. *Mathematical Psychology: An Elementary Introduction.* Englewood Cliffs, NJ: Prentice-Hall, 1970.

Kahneman, Daniel. *Thinking, Fast and Slow.* New York: Farrar, Straus and Giroux, 2011. The Jack and Jill scenario in chapter 9 of the present book is from p. 275 of the hardcover edition.

von Neumann, John, and Oskar Morgen-

stern. *Theory of Games and Economic Behavior.* Princeton, NJ: Princeton University Press, 1944; 2nd ed., 1947.
Savage, Leonard J. *The Foundations of Statistics.* New York: Wiley, 1954.

Chapter 10: The Isolation Effect

Kahneman, Daniel, and Amos Tversky. "Prospect Theory: An Analysis of Decision under Risk." *Econometrica* 47, no. 2 (1979): 263–91.

Chapter 11: The Rules of Undoing

Hobson, J. Allan, and Robert W. McCarley. "The Brain as a Dream State Generator: An Activation-Synthesis Hypothesis of the Dream Process." *American Journal of Psychiatry* 134, no. 12 (1977): 1335–48.
———. "The Neurobiological Origins of Psychoanalytic Dream Theory." *American Journal of Psychiatry* 134, no. 11 (1978): 1211–21.
Kahneman, Daniel. "The Psychology of Possible Worlds." Katz-Newcomb Lecture, April 1979.
Kahneman, Daniel, and Amos Tversky. "The Simulation Heuristic." In *Judgment*

under Uncertainty: Heuristics and Biases, edited by Daniel Kahneman, Paul Slovic, and Amos Tversky, 3–22. Cambridge: Cambridge University Press, 1982.

LeCompte, Tom. "The Disorient Express." *Air & Space,* September 2008, 38–43. http://www.airspaccmag.com/military-aviation/the-disorient-express-474780/.

Tversky, Amos, and Daniel Kahneman. "The Framing of Decisions and the Psychology of Choice." *Science* 211, no. 4481 (1981): 453–58.

Chapter 12: This Cloud of Possibility

Cohen, L. Jonathan. "On the Psychology of Prediction: Whose Is the Fallacy?" *Cognition* 7, no. 4 (1979): 385–407.

———. "Can Human Irrationality Be Experimentally Demonstrated?" *The Behavioral and Brain Sciences* 4, no. 3 (1981): 317–31. Followed by thirty-nine pages of letters, including Persi Diaconis and David Freedman, "The Persistence of Cognitive Illusions: A Rejoinder to L. J. Cohen," 333–34, and a response by Cohen, 331–70.

———. *Knowledge and Language: Selected Essays of L. Jonathan Cohen,* edited by James Logue. Dordrecht, Netherlands:

Springer, 2002.
Gigerenzer, Gerd. "How to Make Cognitive Illusions Disappear: Beyond 'Heuristics and Biases.' " In *European Review of Social Psychology,* Vol. 2, edited by Wolfgang Stroebe and Miles Hewstone, 83–115. Chichester, UK: Wiley, 1991.
———. "On Cognitive Illusions and Rationality." In *Probability and Rationality: Studies on L. Jonathan Cohen's Philosophy of Science,* edited by Ellery Eells and Tomasz Maruszewski, 225–49. Poznan Studies in the Philosophy of the Sciences and the Humanities, Vol. 21. Amsterdam: Rodopi, 1991.
———. "The Bounded Rationality of Probabilistic Mental Models." In *Rationality: Psychological and Philosophical Perspectives,* edited by Ken Manktelow and David Over, 284–313. London: Routledge, 1993.
———. "Why the Distinction between Single-Event Probabilities and Frequencies Is Important for Psychology (and Vice Versa)." In *Subjective Probability,* ed. George Wright and Peter Ayton, 129–61. Chichester, UK: Wiley, 1994.
———. "On Narrow Norms and Vague Heuristics: A Reply to Kahneman and Tversky." *Psychological Review* 103

(1996): 592–96.
———. "Ecological Intelligence: An Adaptation for Frequencies." In *The Evolution of Mind,* edited by Denise Dellarosa Cummins and Colin Allen, 9–29. New York: Oxford University Press, 1998.
Kahneman, Daniel, and Amos Tversky. "Discussion: On the Interpretation of Intuitive Probability: A Reply to Jonathan Cohen." *Cognition* 7, no. 4 (1979): 409–11.
Tversky, Amos, and Daniel Kahneman. "Extensional versus Intuitive Reasoning: The Conjunction Fallacy in Probability Judgment." *Psychological Review* 90, no. 4 (1983): 293–315.
———. "Advances in Prospect Theory." *Journal of Risk and Uncertainty* 5 (1992): 297–323. http://psych.fullerton.edu/mbirnbaum/psych466/articles/tversky_kahneman_jru_92.pdf.
Vranas, Peter B. M. "Gigerenzer's Normative Critique of Kahneman and Tversky." *Cognition* 76 (2000): 179–93.

Coda: Bora-Bora

Redelmeier, Donald A., and Robert J. Tibshirani. "Association between Cellular-Telephone Calls and Motor Vehicle Collisions." *New England Journal of Medicine* 336 (1997): 453–58. http://www.nejm.org/doi/full/10.1056/NEJM199702133360701#t=article.

Thaler, Richard. "Toward a Positive Theory of Consumer Choice." *Journal of Economic Behavior and Organization* 1 (1980): 39–60. http://www.eief.it/butler/files/2009/11/thaler80.pdf.

General

Kazdin, Alan E., ed. *Encyclopedia of Psychology.* 8 vols. Washington, DC: American Psychological Association, and New York: Oxford University Press, 2000.

Murchison, Carl, Gardner Lindzey, et al., eds. *A History of Psychology in Autobiography.* Vols. I–IX. Worcester, MA: Clark University Press, and Washington, DC: American Psychological Association, 1930–2007.

ACKNOWLEDGMENTS

I never know exactly who to thank, or whether to say "whom" to thank. The problem is not a deficit of gratitude but a surplus of debt. I owe so many people that I don't know where to start. But there are people without whom this book simply would not have come to pass, and I'll focus on them.

Danny Kahneman and Barbara Tversky, for starters. When I met Danny, in late 2007, I had no ambition to write a book about him. Once I acquired that ambition, I spent roughly five years making him comfortable with it. Even then he remained, um, circumspect. "I don't think it is possible to describe the two of us without simplifying, without making us too large, and without exaggerating the differences between our characters," he once said. "That's the nature of the task, and I am curious to see how you will deal with it — though not curious enough to want to read it early." Barbara was a different story. Back in the late 1990s, by bizarre coincidence, I taught, or attempted to teach, her son Oren. As I was unaware of the existence of Amos Tversky, I was unaware that he was Amos Tversky's son. Anyway, I went to Barbara bearing a character reference from my former pupil. Barbara gave me access to Amos's papers, and her guidance. Amos's children, Oren, Tal, and Dona, offered a view of Amos that I couldn't have gotten anywhere else. I remain deeply grateful to the Tversky family.

I came to this story as I've come to a lot of stories, as an interloper. Without Maya Bar-Hillel and Daniela Gordon, I would have been lost in Israel. In Israel, over and over again, I had the feeling that the people I was interviewing were not only more interesting than I was but also more capable of explaining what needed to be explained. That this story did not require a writer as much as it did a stenographer. I want to thank several Israelis, in particular, for allowing me to take dictation: Verred Ozer, Avishai Margalit, Varda Liberman, Reuven Gal, Ruma Falk, Ruth Bayit, Eytan and Ruth Sheshinski, Amira and Yeshu Kolodny, Gershon Ben-Shakhar, Samuel Sattath,

Ditsa Pines, and Zur Shapira.

In psychology I was not much more naturally at home than I was in Israel. I needed my guides there, too. For their services in this capacity I'd like to thank Dacher Keltner, Eldar Shafir, and Michael Norton. Many former students and colleagues of Amos and Danny's were both generous with their time and full of insight. I'm especially grateful to Paul Slovic, Rich Gonzalez, Craig Fox, Dale Griffin, and Dale Miller. Steve Glickman offered a lovely guided tour of the history of psychology. And I'm not quite sure what I would have done if Miles Shore had not existed, or had not thought to interview Danny and Amos back in 1983. Miles Shore would be painful to undo.

One way to think of a book is as a series of decisions. I want to thank the people who helped me to make them in this one. Tabitha Soren, Tom Penn, Doug Stumpf, Jacob Weisberg, and Zoe Oliver-Grey read drafts of the manuscript and offered loving advice. Janet Byrne, who will one day be understood as having turned copyediting into an art form, fixed the book so that it was fit for consumption. Without the pushing and prodding of my editor, Starling Lawrence, I wouldn't have bothered to write it in the first place, and if I had, I certainly wouldn't have worked as hard at it as I wound up working. Finally, the possibility that this might be the last book that I ever give Bill Rusin to sell got my rear end in the desk chair sooner than I otherwise would have, so that he might work his magic. But not for the last time, I hope.

The employees of Thorndike Press hope you have enjoyed this Large Print book. All our Thorndike, Wheeler, and Kennebec Large Print titles are designed for easy reading, and all our books are made to last. Other Thorndike Press Large Print books are available at your library, through selected bookstores, or directly from us.

For information about titles, please call:
(800) 223-1244

or visit our Web site at:
http://gale.cengage.com/thorndike

To share your comments, please write:
Publisher
Thorndike Press
10 Water St., Suite 310
Waterville, ME 04901